THE ANGINA MONOLOGUES

Samer Nashef qualified as a doctor at the University of Bristol in 1980 and is a consultant cardiac surgeon at Papworth Hospital in Cambridge. He is a dedicated teacher and communicator, and is recognised as a world-leading expert on risk and quality in surgical care. He is the author of *The Naked Surgeon* and a compiler of cryptic crosswords for *The Guardian* and the *Financial Times*.

THE ANGINA MONOLOGUES

STORIES OF SURGERY FOR BROKEN HEARTS

Samer Nashef

SCRIBE
Melbourne • London

Scribe Publications
2 John St, Clerkenwell, London, WC1N 2ES, United Kingdom
18–20 Edward St, Brunswick, Victoria 3056, Australia

Published by Scribe in 2019

Typeset in Minion by the publishers
Printed and bound in the UK by CPI Group (UK) Ltd, Croydon
CR0 4YY

Scribe Publications is committed to the sustainable use of natural resources and the use of paper products made responsibly from those resources.

9781911617785 (UK edition)
9781925713817 (Australian edition)
9781925693416 (e-book)

CiP records for this title are available from the British Library and the National Library of Australia.

scribepublications.co.uk
scribepublications.com.au

CONTENTS

Foreword 1

1 Conflict of interest 5
2 Heart surgery for beginners 15
3 Getting on with it 27
4 Mild paranoia 35
5 CABG and how to avoid it 47
6 An easy cabbage 67
7 World-class surgery on a shoestring 91
8 The story of A and H 105
9 When the pump is broken 113
10 Sabotage 131
11 Irony 155
12 McKlusky's 167
13 Will *you* do the operation, Doctor? 185
14 Brazilian 191
15 Keyhole surgery and other novelties 205
16 This time it's personal 219
17 The many forms of Lazarus 231
18 An accidental fraud 243

Appendix: A crossword puzzle for Roger Whiting 271
Acknowledgements 275

angina

Pronunciation: /anˈdʒʌɪnə/

NOUN

(also angina pectoris /ˈpɛkt(ə)rɪs/)

A condition marked by severe pain in the chest, often also spreading to the shoulders, arms, and neck, owing to an inadequate blood supply to the heart:

'he had high blood pressure and he suffered from angina'

Latin *pectoris* 'of the chest'

Oxford English Dictionary

To all the patients I have been privileged to treat

FOREWORD

Sometimes even dictionaries can get it wrong.

Angina is not a pain. It is more of a discomfort or pressure, a tightness or weight. And it is not usually severe, but is often accompanied by a feeling of impending doom, which makes you stop whatever it is you are doing in the way of physical exertion until it abates a few minutes later. It is also the Number One symptom that drives heart patients to seek medical help, with Number Two being shortness of breath.

This book relates the real-life stories of patients — with angina and without — who had major open heart operations to fix their cardiac problems. It tells of the triumphs and disasters of heart surgery, of the places where it is done and the professionals who do it, but always with a focus on the people who are at the centre of it all: the patients.

In writing this book I have selected stories that I hope provide a vivid illustration of different heart surgery patients, a wide range of heart conditions and the variety of procedures that a surgeon undertakes. Some of these stories are gory, some are funny and some have happy endings. Some of them have taken place at a time when I was able to put pen to paper (or fingers to keyboard) immediately, and, when that happened, I wrote down the events as they unfolded. Most of the other stories I have selected from memory and I have chosen them because they are memorable. Necessarily, some of these do not have a happy ending and that is precisely the sad reason why they are memorable. The aim of this book is, however, neither to scaremonger nor to sensationalise, but to provide an honest insight into my world. This is of course a world of drama, of life and death. It is also a world in which some of the finest attributes of human nature – inventiveness and resilience – shine most brightly both in the patients and in the people who care for them. Finally, it is a world which actually matters directly to you because, more likely than not, one day, either you or someone you love will need a heart operation. I hope that you will find this insight illuminating as well as interesting.

Few areas of human endeavour are laden with as much drama as heart surgery. With death always a presence and sometimes an immediate threat, the situations faced by patients and those who care for them can be emotional and nerve-wracking. People who deal with such stressful situations often evolve a number of coping mechanisms. Prominent amongst these is a dark and uncompromising sense of humour, one which probably exists in many surgical specialties, particularly those at the sharper end of the field. Some of these humorous and highly irreverent instances appear here, and I should like to apologise in advance if anyone finds such narrative offensive. I believe that I speak for all my colleagues when I say that such dark humour is, to us, merely a way of dealing with what sometimes can be unbearable stress. The fact that laughing in the face of adversity helps us through our working lives in no way diminishes how passionately we care about our patients and our work.

CHAPTER 1

Conflict of interest

Sometimes, in the words of Bob Geldof, I don't like Mondays, and yesterday was just such a Monday.

It was the first working day in January after the long Christmas break and, as such, it was something of a shock to the system. Instead of a gentle start, easing back in to work, I had two big operations on the first day. They were both complex, dangerous and challenging.

The first patient was a 47-year-old man. He needed two procedures done on his heart. Neither of the two was particularly challenging in itself, but the combination of them together in one operation was something I had never done before. The first procedure was a pericardiectomy and the second was a mitral valve repair. The first simply means removal of the pericardium or, in plain words, stripping the heart of its outside lining. This lining is normally a smooth and slippery bag, with plenty of room for the heart to move within it, and a little bit of fluid in which the heart can beat freely. This patient's pericardium was no longer a smooth bag with a bit of fluid. The fluid had long gone as his pericardium had become thickened, scarred and rigid, firmly stuck to the heart within it and shrunken, strangling the heart, as in a straitjacket. This restricted the pumping action of the heart and caused heart failure. The second procedure was to repair his leaky mitral valve in the middle of the heart, which was making his heart failure even worse.

The trouble is that these two conditions do not usually come together and the two operations needed to fix them do not agree with each other. The first should ideally be done without using the heart-lung machine — to reduce the risk of bleeding from the raw area after the lining is stripped off the surface of the heart — and the second simply cannot be done without the heart-lung machine, thus greatly increasing the risk of bleeding from the first. To make matters worse, during the operation there was one bit at the back of the heart where the lining was so calcified and stuck that it proved impossible, despite many attempts, to separate it from the heart muscle without tearing the heart to shreds, so that the heart remained stuck at that point, making access to the mitral valve very difficult. Without being able to see the valve properly, I ended up repairing it mostly by feel. Thankfully, it worked, but this was more by luck than by judgement or skill.

The second patient, a grand, 79-year-old man, was the retired chairman of the board of a nearby hospital. He had sought treatment at Papworth as he was considered neither fit enough nor young enough to have such a complex operation locally. He needed a quadruple heart procedure: an aortic valve replacement, a double coronary bypass, a hole in the heart closed and a 'maze' operation to correct an irregular heartbeat. This would be really pushing the limits in a 40-year-old, let alone a patient approaching his eightieth birthday. At his age, nobody would have been unduly surprised if his elderly body struggled to cope with such a heavy surgical assault.

Fortunately, both operations went well. At home, just before going to bed at about midnight, I made one final phone call to the intensive care unit (ICU) to check on the two patients, and was assured that both were stable and progressing well. As I woke up the following Tuesday, the first happy thought that crossed my mind was that there had been no phone calls from the ICU during the night — a good sign!

A bright winter sun was shining in a cloudless sky and I briefly considered riding the motorbike to work. I immediately dismissed the idea as daft as soon as I stepped outside, felt the bitter cold of that January morning and sensibly decided to take the car. Driving the car to work provides the added double advantage of a cup of good black coffee on the way to the hospital and the ability to listen to the *Today* programme on BBC Radio 4 to catch up with what is happening in the outside world.

On that particular Tuesday morning the *Today* programme was reporting that Israeli forces had widened their attacks in the Gaza Strip after heavy fighting, with disparate claims and counterclaims by each side in the conflict on the numbers of soldiers and civilians killed. I remembered that, two years previously, I had been asked by a charity to help set up a heart surgery service in the Gaza strip, and I had declined in view of the volatile situation. I had felt a bit of a coward at the time and volunteered my services instead to the much less dangerous West Bank.

I parked the car at the hospital and walked the short distance to my office. This took me along the border of the famous Papworth Hospital duck pond, a circular body of water about 100 metres across with a small island in the centre of it. The pond was now mostly frozen, but the resident ducks were nevertheless still quacking happily, despite being confined to a small crescent of still liquid but near-freezing water towards the edge. At the time, I was the Chair of the consultants' committee at Papworth Hospital and, a few years previously at one of our monthly meetings, one consultant colleague, who was a chest physician, had asked a pertinent question about whether the duck pond posed an infection risk to our chest and transplant patients. I referred the matter to the consultant microbiologist. She then stood up and addressed the assembled group: 'Who here wants to keep the duck pond?' All hands went up. 'In that case,' she continued, 'do not ever ask me that question again!'

I walked into my office, which had a large window providing a fine view of the said duck pond, switched on my three computers. I used three in those days for the simple reason that information technology in the NHS is relatively slow and the machines are out-dated. This means that every command takes a machine a few seconds at least to deliver the goods. With three computers, while one of them is thinking about opening a file, I can move on to the next to do something else. Even a few seconds saved here and there will help in making me more efficient. One computer is my clinical patient database exclusively. One allows me to see all the images of medical investigations. The third is for email and everything else (including crosswords). Since then, things have moved on, but not, sadly, the quality of the computers. I now use four.

I reviewed the tests on the day's patients in preparation for surgery, quickly checked for any urgent emails requiring an immediate response and went to change into scrubs. Our male changing room is small, windowless and utterly chaotic. It is stuffed with banks of lockers and its floor is always haphazardly strewn with operating theatre shoes and discarded scrubs. To make matters worse, somebody in the department must think it funny to empty the rubbish from his pockets into other people's theatre shoes and I had, in the past, found all manner of detritus in mine. On this occasion I found a pair of disposable scissors, a sweetie wrapper and a slip of paper with the results of a blood test. I removed them and, just before I threw them away in the dustbin, I recognised the name of the person who had requested the blood test – it was one of the ICU nurses on duty the previous night. I wondered briefly about confronting him about this bizarre antisocial behaviour, but promptly forgot all about it.

Before going to the operating theatre I paid a very quick visit to the ICU to see yesterday's patients. They were both, to my relief, looking very well indeed. The younger man had not bled after all, had made a rapid recovery and was awake and having his breakfast. The older man was still a little drowsy, but looked far better than could have been expected after my massive surgical onslaught on him the previous day, so it was with a light heart and a spring in my step that I went to the operating theatre to start the morning case. It was an operation that is a pure joy to do: a single coronary artery bypass graft (or CABG) in an otherwise fit and healthy patient.

One feature of heart surgery, when compared to some other surgical specialties, is that there is no 'small fry': a single coronary bypass is about as close as we heart surgeons can get to a simple, straightforward operation, and even that can be fraught with hazard, although, in comparison with the previous day, this would be a breeze. I resolved to assist Betsy Evans, my then registrar, in performing the procedure and, while she was setting up the case, I went to the theatre dining room for coffee, banter and a glance at the cryptic crossword. It was going to be a good day. The only cloud on the horizon was that I was on call for emergencies but, much of the time, very little happens on that front.

Betsy did a superb job in the single CABG. We were finishing and tidying up in preparation for closing the chest when another registrar came into the operating room. He, too, was on call for emergencies that day, and he informed me that we had just been referred a 39-year-old woman from Norfolk with a confirmed diagnosis of acute aortic dissection. She was already on her way to us from Norwich, some 90 miles away, in an ambulance with the blue lights flashing.

Acute aortic dissection is possibly the only real emergency in heart surgery. Most urgent heart conditions can be made less urgent with drugs and devices, so that the operations needed to fix them can then be carried out in a safer and more-or-less planned manner a day or two later. Acute aortic dissection cannot be treated this way: it demands surgery, and demands it immediately.

This is what happens: the inner lining of the aorta — the biggest artery in the body — is suddenly torn because of weakness or high blood pressure or both. The patient experiences a sudden, searing chest pain that shoots down the back. The pain is so severe that the patient sometimes collapses as a result. Meanwhile, the highly pressurised blood within the aorta is seeping into the tear and advancing between the layers of the wall of the aorta, peeling it off like badly applied wallpaper: in this manner, the blood 'dissects' the wall of the aorta. In doing so, it travels backwards towards the heart, where it can disrupt the aortic valve, making it leak. It can also shear off the coronary arteries, producing a heart attack; and, travelling forwards, it can threaten to block or disrupt any artery that comes off the aorta, which is, essentially, all of them. The heart, the brain and every single organ in the entire body are put at risk in acute aortic dissection, and on top of all of that, the aorta itself may rupture, causing instant death by massive bleeding. In the first two or three days of acute aortic dissection, the death rate is 1 per cent every hour, so that this is one condition where there is no time to lose.

This particular patient, however, had a further complicating feature: she was 37 weeks pregnant, and with twins.

If yesterday's patient had an in-built conflict between the best way to approach the lining of his heart and that for his mitral valve, then Nina, the pregnant woman with acute aortic dissection, had a worse conflict, magnified several times. Both of yesterday's patients, their trials and tribulations and any preoccupation I had over them immediately went out of the window. This situation demanded immediate and intense concentration.

Nina's best chance of survival would be secured by keeping her blood pressure really low, until an immediate operation repaired her acute aortic dissection. Nina's twin babies, however, may not survive their mother being put on a heart-lung machine, and they needed continuous good blood pressure to supply the placenta and keep their little bodies going. All three – the mother and her unborn twins – were in grave danger. Whose interests should we put first?

The on-call anaesthetist John Kneeshaw and I hastily arranged a makeshift case conference in a little side room on the ICU. We considered all of the options, and consulted the obstetricians and neonatologists (newborn-baby specialists) at the nearby Addenbrooke's Hospital in Cambridge. They told us that as far as they were concerned, 37 weeks is not far off a full term in pregnancy, and that they were confident that — if the babies were delivered now — their chances of survival would be excellent. We immediately dismissed the option of inducing a normal labour: the high blood pressure that would be caused by the pain of contractions would almost certainly burst Nina's damaged aorta. After brief consideration, we also dismissed the option of going ahead with repairing the dissection and letting the babies take their chances of survival: it seemed so unfair when they were able to survive outside the womb already. Only one option remained: a rapid Caesarean section, under general anaesthetic, with immaculate blood pressure control, to be followed by a brief and somewhat impatient wait for the afterbirth and for the womb to shrink down to reduce the risk of massive haemorrhage from the raw area when the heart operation was begun. We worked out that this should delay the heart operation only by an hour or two at the most, an increased risk to Nina's life of no more than 2 per cent, which we thought was just about acceptable under the circumstances. (I am aware of the brutality of this sentence: a 2 per cent risk to a young woman's life being seen as 'acceptable' is shocking, but that is one dehumanising aspect of having to deal with situations where life is at risk and the best we can do is choose the least risky of several perilous paths.)

The ambulance then arrived. Nina had a rapid assessment, followed by a quick chat to John and me, in which we outlined our plan. She was in dire straits: cold, sweaty and in shock. Her aortic valve had already been disrupted by the dissection and was leaking badly. As a result of the leak, her heart was failing and she was desperately short of breath. She readily agreed to our proposed plan, shakily signed the consent form, and we whisked her into the operating theatre.

The obstetricians from Cambridge came to Papworth, accompanied by the neonatologists with two incubators in tow, ready for the new arrivals. John put her to sleep and set up a combination of powerful intravenous drugs to control her blood pressure, allowing him to tweak it either up or down as required. The obstetricians scrubbed up, carried out the Caesarean section and delivered the babies, who started breathing immediately and looked perfectly ready to face the world. While the obstetricians were sewing up the wound, the babies were brought out of the operating theatre in their respective incubators.

Babies born by Caesarean section do not suffer the trauma of going through the birth canal, so their faces do not become scrunched up and bruised in the process, and they do not acquire the strangely wizened old-person look that many babies have when they first meet the world. Alfie and Evie looked simply gorgeous: a boy and a girl with beautiful, blue, wide-open eyes, breathing comfortably and without a care in the world. I was looking at them in their cots in wonder, when John came out of the operating room. 'Yeah, they're really cute, but that's enough cooing over them,' he said. 'Now bloody get in there and make sure they still have a mother.'

CHAPTER 2

Heart surgery for beginners

Before I can tell you any more about what happened to Nina and her heart, I should like to pause briefly to go over the structure and function of our most vital of vital organs. Since this book will tell stories of the heart, it seems to me that it must be helpful to the non-specialist reader to have a basic understanding of what it is, how it is plumbed, what can go wrong with it and how it can be fixed. Reassuringly, the heart is nowhere near as complicated as most people might think it is, and a little basic knowledge of its workings will help make sense of much of the content of this book.

Despite centuries of poetry, literature and romance, the heart is not the seat of emotions. It does not hold an individual's personality within its walls and it looks nothing like its shiny and rosy depiction on valentine cards. It is, very simply and prosaically, a pump. Its job is to push the blood around the body. It is actually two pumps in one. The right side of the heart receives the used-up, blue blood that has already delivered its oxygen to the body and pumps it to the lungs, where it picks up fresh oxygen. The now pink, oxygenated blood then comes back to the left side of the heart, where it gets pumped to the body to deliver oxygen and goodness. Then it comes back to the right side for the whole process to start again.

This is how blood circulates through the heart and lungs: starting on the right side, two huge veins (each a little wider than a thumb) bring the used-up depleted blue blood from the body back to the heart. The veins enter the right atrium, which is the collecting chamber of the right heart, from where blood travels to the right ventricle, which is the actual pumping chamber. The right ventricle then pumps the blood to the lungs through a big artery called the pulmonary artery. In the lungs the blood dumps carbon dioxide, picks up oxygen and comes back to the left atrium through four big pulmonary veins. The left atrium is the collecting chamber of the left heart, from where blood enters the left ventricle, the actual pumping chamber. The left ventricle then pumps the oxygen-rich pink blood into the aorta, another big artery with branches to deliver blood to every organ in the whole body.

The design of the heart as a pump is very simple: it is a bag made of muscle. When it is full of blood, the muscle squeezes (that's exactly what a heartbeat is) and the blood comes out. To make sure that the blood travels forward, not backwards, there are four valves. Two sit at the entrance and exit of the right ventricle, and two sit at the entrance and exit of the left ventricle. These valves allow flow in one direction only when the ventricle begins to squeeze, so that blood does not flow backwards. At the entrance to the right ventricle is the tricuspid valve, so called because it has three cusps. At the exit is the pulmonary valve, so called because it leads to the pulmonary artery. On the left side, at the entrance to the left ventricle is the mitral valve, so called because it supposedly resembles a bishop's mitre, and, at the exit, the aortic valve, because (you guessed it) it leads to the aorta.

In simplified graphic form, the right heart can be shown like this, with arrows indicating the direction of blood flow.

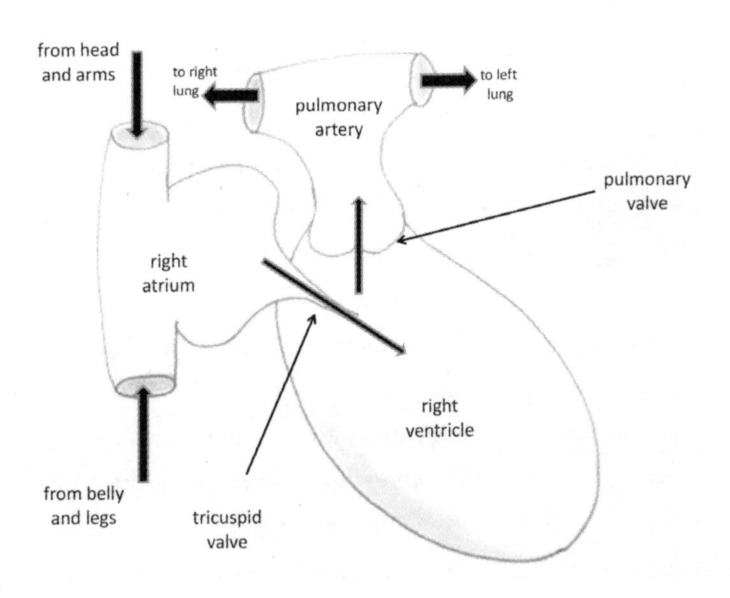

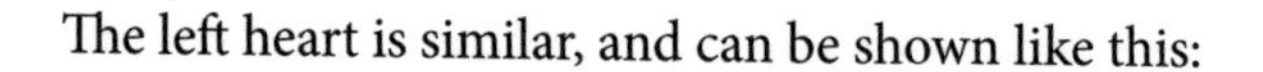
The left heart is similar, and can be shown like this:

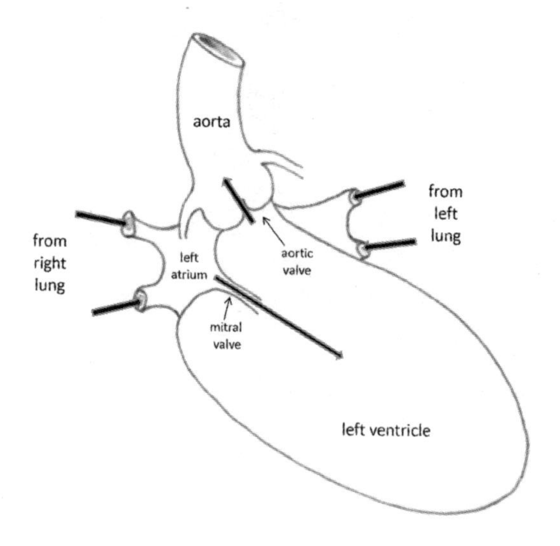

One last thing about the left heart needs to be mentioned: right at the very beginning of the aorta — just after the aortic valve — you can see two branches: two little arteries coming out of the aorta and heading back towards the heart: these are the famous coronary arteries. Their job is to supply blood and oxygen to the heart muscle itself.

Another little-known fact about the heart is that the anatomists have somewhat misnamed its two principal parts. The right heart is much more towards the front than the right side, and the left heart is much more towards the back. When we put the two together, the right heart sits mostly in front of the left heart, and the whole heart actually looks something like this:

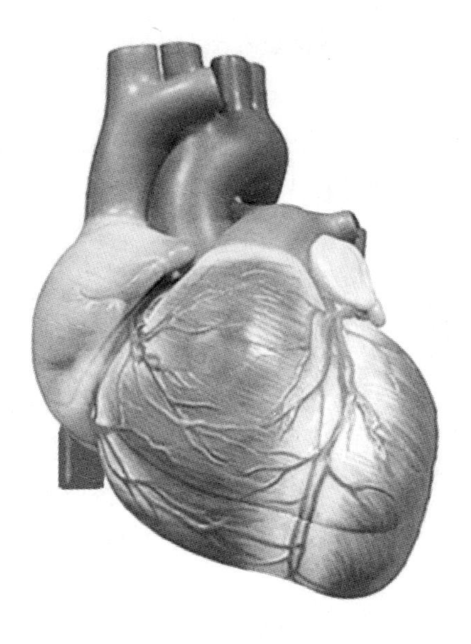

Heart surgery is often described in sensationalist terms: 'a miracle of modern medicine', 'a triumph of technology' and so forth, but what is amazing about heart surgery is not that it exists and actually works, but that it took an inordinately long time to make an appearance on the medical stage. Medicine, in its myriad forms, has been in existence for thousands of years. The Hippocratic oath was written around 2,500 years ago. Anaesthesia has existed since the nineteenth century and all types of operations have been done under general anaesthetic since then, yet heart surgery only began its first faltering steps in the late 1950s, barely 60 years ago.

What kept it so long? After all, the heart is a pump, pure and simple. When something goes wrong with a pump it is a plumbing problem, needing technical plumbing solutions. How else do you fix a blockage in a pipe or a leaky valve? By swallowing a tablet? Yet for more than two millennia the heart remained exclusively the domain of the physician, with the surgeon firmly excluded from the scene. The best that a heart patient could hope for was pills and potions, while the narrowings, blockages and leaks were left untouched by medical hands. In fact, the taboo on operating on the heart was so strong that Dr Theodor Billroth, one of the great founding fathers of modern surgery, stated in 1889 that 'a surgeon who tries to suture a heart wound deserves to lose the esteem of his colleagues'.

There were two main reasons why the heart could simply not be tackled surgically until some clever devices finally became available.

The first reason actually had to do with the lungs. These are the sponge-like organs that exchange gases with the air. They inflate and deflate about a dozen times per minute to bring in oxygen and get rid of carbon dioxide. The problem is that the lungs do not do this by themselves, because they are entirely passive structures: they have no muscle and cannot move. The lungs inflate and deflate only by following the chest wall around them as it expands and contracts with the muscles of breathing. There is a sealed vacuum between the lungs and the chest wall, so that the lungs must follow the movements of the chest wall with every breath. Opening the chest to operate on anything inside it will break that seal and let air into the chest cavity. The lungs then fall away from the chest wall and breathing stops immediately. At first, our intrepid pioneering heart surgeon would have been pleased to see this: there is suddenly a lot of room in the chest and the heart can be easily reached. Sadly, his joy would have been short-lived, because the patient would have died a few minutes later from lack of oxygen.

This was the fate of patients in whom chest surgery was attempted until the second half of the nineteenth century, when the endotracheal tube was invented. By inserting this tube into the windpipe (trachea), air or oxygen could be actively blown into the lungs, regardless of whether the chest cavity was closed or open. This of course made a lot of anaesthesia for all kinds of surgery safer and more controlled: the patient can be put into such a deep sleep so as to stop breathing, and the anaesthetist can take over the breathing function by blowing air or oxygen directly into the lungs through the endotracheal tube. It also made major, open-chest operations possible for the first time.

The second reason was the heart itself. This little muscle, about the size of your fist, pumps five litres of blood every single minute to deliver oxygen and nourishment to the entire body. In fact, the average adult has only about five litres of blood in total, so your entire lifeblood goes full circle around your whole body every single minute of your life. If the heart stops, death follows almost immediately, as the body cannot live without a supply of blood and oxygen. Different parts of the body, however, are not equally sensitive to the loss of their blood supply for a short while. Your leg will probably recover fully if its blood supply is cut for half an hour, but your brain most certainly will not survive more than a few minutes without blood and oxygen under normal circumstances.

Operating on the heart involves touching it, twisting it, pressing on it and sometimes turning it upside down. All of these manoeuvres interfere with the pumping action, so any heart operation that disrupts the pumping action of the heart for more than a few minutes is likely to cause brain damage or death. Thus, the only heart operations that could be done in the past were ultra-short ones: a few minutes to cobble up a hole and hope for the best marked the limit of what heart surgeons could do. This unhappy state of affairs remained until the middle of the twentieth century, when the heart-lung machine was invented: a contraption that took over the job of the heart and lungs, keeping the patient alive while surgeons fiddled with the heart. After the first successful use of the machine — in 1953 by John Gibbon in Philadelphia, USA — everything changed: this marvellous invention opened the door wide and ushered in the new specialty of cardiac surgery.

The invention of the heart-lung machine was to heart surgery as the starter pistol is to an Olympic sprinter. The specialty took off and ran with breathtaking, almost indecent, speed, so that by the 1960s heart surgery was no longer considered crazy. More and more patients were being saved, more sophisticated and complex operations were being invented, and the results were looking better and better. The specialty was rapidly transformed from a very limited, last-ditch intervention in otherwise hopeless cases to a routine and substantial part of modern medicine's armamentarium.

Putting a patient on the heart-lung machine is relatively simple. First, we give the patient a huge dose of heparin, a drug that prevents clotting so that the blood remains liquid, despite touching the plastic and other artificial surfaces of the circuit of the heart-lung machine. Then we have to catch the blue blood (that has already delivered its oxygen) before it reaches the heart and lungs. To do that we place a large tube usually in the right atrium. This tube drains the patient's blood into the heart-lung machine, where the blood gets oxygenated (the lung part) and pumped (the heart part) into another tube which we normally place in the aorta, so that it delivers blood to every artery in the body.

Once the pipes are in place, we switch on the machine and set it to pump about five litres per minute, which is what the heart usually does. While the machine is running, we can do whatever we like to the heart and the patient still stays alive. Two interesting things happen when the heart-lung machine is switched on. The first is that the pulse disappears. The only reason we have a pulse is because the heart pumps in beats: it squeezes and relaxes, and every squeeze can be felt as a heartbeat in your chest or as a pulse in your wrist or anywhere else where an artery is close to the skin. The pump of the heart-lung machine does not do this: it pumps continuously, so that the patient having a heart operation is fully alive, but without a pulse. The second is that the lungs are now completely inactive: there is no more blood going to them to pick up oxygen, as all the blood is diverted to the heart-lung machine before entering the right heart, and it picks up oxygen in the heart-lung machine. This means that there is no point whatsoever in the patient breathing and the anaesthetist usually switches off the ventilator. The patient is still fully alive, not only without a pulse, but also without breathing.

At the beginning of a standard heart operation, the surgeon starts by opening the chest using a power saw to cut in half the breastbone (or sternum). A retractor is used to spread open the chest and the surgeon opens the pericardium (the bag around the heart). The surgeon then places the two big pipes connecting the patient to the heart-lung machine: one pipe goes in the right atrium to capture the blue blood and one in the aorta to deliver the pink blood. The machine is then switched on, and the operation can now begin, but not quite. When the heart-lung machine takes over the work of the heart and lungs, the heart is no longer working, but it is still beating, albeit empty. This presents a moving target, which is not ideal for any surgery, especially if the procedure is intricate. Worse than that, however, is that many areas of the heart are still not accessible. To change an aortic valve, for example, a large cut needs to be made in the aorta, and when that is done, the five litres a minute that the heart-lung machine is pumping will end up on the floor. Doing a CABG operation involves making a hole in a coronary artery. Of course, a coronary artery is not big enough to bleed at five litres a minute, but it will bleed at about 50 to 100 millilitres a minute, enough to spurt the surgeons in the eye and prevent them from seeing what they are doing. To avoid all of these things happening, we place a clamp on the aorta, isolating the heart from the circulation provided by the heart-lung machine. Now we can open any chamber in the heart and, once the blood that is already in there is sucked out, no more will come. Problem solved! When I explain this to smart medical students, they immediately exclaim: 'But what about the heart itself? It, too, needs blood and oxygen to survive!'

They are, of course, absolutely right. As soon as the clamp goes on the aorta, the heart starts to die slowly. The only thing we can do is to slow the process of dying even more by infusing it with a cold solution containing potassium. The cold slows everything down, allowing the heart to survive much longer without oxygen, and the potassium paralyses it, making it unable to beat at all, and thus dramatically reducing its need for any oxygen. Under these circumstances, with a patient who is asleep, a machine doing all the work of the heart and lungs, and a cold, still heart, we have about an hour or so to complete the heart operation with no adverse effects on the heart itself. If the operation on the heart takes longer, we need to do a few more things, such as cooling it further, topping up the potassium solution and so forth. If it takes more than four hours, the heart will have been damaged. This is the reason that when the clamp goes on to the aorta, a clock starts ticking and there is a noticeable and abrupt change in the atmosphere in the operating theatre: idle chit-chat stops, movements become faster and the entire team concentrates intently on the job at hand. The less time the aorta is clamped, the better it is for the heart and, by extension, for the patient.

In summary, these are the preparatory steps for a heart operation:

- Open the chest
- Give an overdose of heparin to prevent any clotting
- Put a pipe in the aorta and a pipe in the right atrium and connect them to the heart-lung machine
- Start the heart-lung machine
- Clamp the aorta
- Give the potassium solution
- Get on with it

CHAPTER 3

Getting on with it

Nina's Caesarean section wound was being dressed while Betsy and I were washing our hands. The afterbirth had been delivered and we had already waited for some 20 minutes to make sure the womb had shrunk down and that its raw lining was no longer bleeding, or at least no longer bleeding profusely. John had Nina's blood pressure perfectly under control and, for now at least, things looked as stable as they could be in the circumstances. The surgical drapes used in the Caesarean section operation were removed. Nina's skin was prepared again with antiseptic solution, and new surgical drapes were applied, exposing only the chest and the legs. The reason for keeping the legs accessible is that, if there is a problem with a coronary artery, a piece of vein may be needed in a hurry, and the lower leg is the best place for finding one.

Once a patient is covered in surgical drapes, and all that

can be seen of her are the strips of skin where the incisions take place, she becomes — I am reluctant to say — somewhat less of a human being in the eyes of the surgical team, and more of a technical problem needing a technical solution. This is not necessarily a bad thing. Surgeons are at their best when they provide the best technical solution possible in a particular situation. The fact that she is young, has two newborn babies, works for the fire and rescue service and has an adoring husband in the armed forces are important during the planning and decision-making phases of the treatment. Once that is complete, a course of action is selected and the operation begins, these facts are, at best, a distraction and at worst, information which may cloud the fine technical judgement that is needed in the context of the operating theatre. Thus I fully and gratefully recognised the palpable loss of my holistic approach to the patient as I made the incision. Until that point, I had viewed this particular operation with quite a bit of trepidation, but no more. Nina was no longer Nina, but a dissected aorta needing repair as safely and as expeditiously as possible.

Which operation we would do very much depended upon what we found inside. At the very least, the ascending aorta above the coronary arteries would have to be replaced, and the aortic valve, which we knew had crumpled as a result of the aortic dissection, somehow tacked back or 're-suspended', so that it would work as a valve again. This would be the 'best-case scenario', but if the coronary arteries had been torn off by the dissection, and if the root of the aorta, where the coronaries emerge, was so damaged that it, too, had to be replaced, then the root of the aorta would have to be changed as well, meaning that the coronary arteries would have to be detached and re-implanted into the new aorta and the valve would almost certainly need to be replaced in addition. The worst-case scenario was if the tear in the aorta spread into the next bit, which is the aortic arch from where the blood vessels to the brain arise. If that part of the aorta also needed to be replaced, then it would become a very big operation indeed and the risk to Nina's brain would become substantial.

I opened the pericardium, the smooth and shiny bag in which the heart lives. The dissected aorta was clearly and vividly visible. A normal aorta is a cream-coloured tube, two to three centimetres in diameter. This aorta was a huge, angry-looking, bloated red sausage, with swirling blood seen clearly where it should not be, through its gossamer-thin outside wall, which was all that stood between life and death. Should that flimsy layer give way, there would be instant massive haemorrhage. The diagnosis was confirmed. The ugly, bloated and purple aorta, teetering on the edge of rupture, stood in sharp contrast to the heart itself, a young, pink and willing organ, beating quite happily and with a certain elegance to its rhythmic contraction, almost like a dance movement with a final flourish at the end of every beat.

A dissected aorta cannot be used to join the patient to the heart-lung machine, because it is shredded to bits and will need to come out anyway, and so we had to find another artery. Fortunately, the system of arteries that feed the body is like a grid where every artery is connected to every other, so that pumping five litres a minute into any artery will reach every other artery and the whole body will be supplied with blood and oxygen, as long as the artery chosen for access is reasonably big, so that it can handle such a large amount of blood flow (the artery in your little finger won't do). We made another cut over the axillary artery as it passes under the collarbone and prepared it for this purpose, joining it with a pipe to the heart-lung machine. After that, taking care to stay well away from the delicate layer which was just managing to hold the blood within the aorta, we joined another pipe to the right atrium and started the heart-lung machine. The blood flowed, blue and devoid of oxygen, from the right atrium into the machine and came back, bright pink and laden with oxygen, into the axillary artery, where it flowed to every artery in the body. 'Full flow,' the perfusionist announced to indicate that we had reached the five litres or so per minute. 'Lungs off,' said John as he switched off the ventilator. The operation to fix the aorta could now begin.

I asked the perfusionist temporarily to drop the flow from the heart-lung machine, so that I could safely apply a clamp to the tattered aorta without disrupting it further. 'Down on flow,' I requested. The usual salacious rejoinder from this particular perfusionist ('Lucky old Flo') was not forthcoming – instead she simply said: 'Flow down.' There was an uncharacteristic sobriety in the ambience of the theatre that day, reflecting the seriousness of the situation.

I gingerly applied a soft, padded clamp right across the aorta, as high as possible but just short of the arteries to the brain. The perfusionist then raised the flow to the normal level and, from this point onwards, all the organs in the body were being supplied with blood and oxygen, except for the heart itself, which was now isolated from the circulation and receiving no oxygen. As soon as the clamp is applied, the heart starts to die and the time has come to 'get on with it'. We gave the potassium to the heart and it stopped beating within seconds. Nina's life now depended entirely on the flow of blood supplied from the heart-lung machine until her heart started beating again. I cut open the aorta, sucked out any blood that was within it and removed the clots that had formed between the layers of its wall.

I then proceeded to inspect the damage. There it was: the original breach in the lining of the aorta was in the root of the aorta just above the valve, but, thankfully, well away from both coronary arteries, so that replacing the entire aortic root would not be needed. I cut out the ascending aorta and also the part of the aortic root containing the tear. I used a powerful surgical glue to stick the separated layers of the aorta back together. I then used three reinforced stitches to anchor the aortic valve back to where it was before it collapsed from the acute dissection. I checked that the valve 'looked right' and worked well.

We then took, from the shelf, an artificial aorta of the correct size. This is essentially a white tube made out of woven Dacron. I trimmed it to a shape that would replace the ascending aorta, with a 'tongue' that would replace the bit of the aortic root I had also cut out. I then stitched in this graft to replace the aorta. This needed two suture lines: one near the heart just above the valve, and one just before the take-off of the brain arteries.

Finally, when the artificial aorta was securely in place, we washed out the air that had entered the heart, as any air left in the heart when it rejoins the circulation will be pumped out when the heart resumes beating and could cause mischief anywhere in the body, depending where it ends up. If it reaches the brain, a stroke may be the result. Washing the air out is never an elegant manoeuvre. The perfusionist running the heart-lung machine allows some blood to enter the heart, and we use that blood to rinse the air out by shaking the heart, squeezing and generally wobbling it about until no more air bubbles are visible. Getting rid of the air is also a sign that the operation is nearly over and the agitation that accompanies it is usually sufficient to wake up the sleepiest anaesthetist from deepest torpor by signalling that the end of the operation is finally in sight.

The repair was now over and the clamp on the aorta was removed. The whole thing had taken 72 minutes, an acceptable time as far as the heart was concerned. As soon as the clamp was removed, blood flowed down the coronary arteries and washed out the potassium solution, so the heart sprang into action. We gave it a few minutes to recover from the oxygen deprivation that we had inflicted upon it, and it certainly looked ready to take up its pumping duties once again. John switched on the lungs and we disconnected the heart-lung machine.

Nina's heart worked perfectly. I left Betsy to close the chest and walked out of the operating theatre with a feeling of massive relief that everything had gone as planned. Barring unforeseen complications, Nina was going to be fine and Alfie and Evie would have a mum after all.

CHAPTER 4

Mild paranoia

Heart surgery can do unpleasant things to the brain. Whenever we propose a major heart operation to a patient, we of course always seek fully informed consent. This involves telling the patients what the operation will do for them in terms of benefits, such as relieving symptoms like chest pain and breathlessness, and helping them avoid dangerous events in the future, such as heart attacks, heart failure and death. We also inform our patients of the likely major risks of such a major operation, such as death from the operation itself. Fortunately, the risk of death from a heart operation is now quite small, at less than 2 per cent overall, but for individual patients the risk varies enormously from less than half of 1 per cent to more than 90 per cent. This of course depends on how old and sick the patients are, what state their heart is in and how big an operation is being contemplated.

Most heart surgeons, when discussing the risk of an

operation, will also cite a small risk of stroke, of around 1 per cent or less for the majority of patients. The reasons that the catastrophic complication of stroke is an ever-present risk in heart surgery are legion. Some of these reasons have to do with the operation itself. Manipulating the aorta upstream from the blood vessels to the brain can dislodge tiny fragments of atheroma (the 'cheese' that coats blood vessels as we age) and these can fly off to the brain. Putting the patient on the heart-lung machine changes the dynamics of how blood flows to organs, including the brain. Opening the heart for, say, a valve replacement or repair lets air into the heart, and it can be difficult to ensure that every last bubble is removed before the heart is reconnected with the circulation, and air in the brain can also cause a stroke. Finally, patients with coronary artery disease often have furred-up arteries elsewhere, including the brain arteries, and are at risk of a stroke anyway. All of this means that the risk of stroke, although quite small, is forever present.

If an obvious, major stroke is fortunately rare, less serious brain upsets are very common. Patients often find that they are a little confused in the first few days after an operation, that their brain function is not as sharp as it used to be and that their ability to concentrate has suffered. Quite commonly they can also develop mild psychiatric disturbances, such as paranoia, hallucinations and disorientation in time and space. Fortunately, most of these effects are transient and relatively easy to deal with, but sometimes they can be troublesome or embarrassing.

I remember one extremely prim and proper old lady who had a straightforward heart operation and who, for a few days afterwards, would get up in the middle of the night and try to climb into bed with other (male) patients on the same ward. The effect did not last long and she reverted to her prim and proper old self again by the time she went home a week later, but she remembered with alarming clarity what she did and was, of course, absolutely mortified about her behaviour. It took a long time to reassure her, during the post-operative follow-up appointment, that this really was not her fault, and that she truly was not a naughty person, and that she could blame the surgery entirely for her being so out of character.

More common than this loss of social inhibition is the development of paranoid delusions, where patients become convinced that there is an elaborate and malevolent conspiracy against them in which the entire surgical care team is involved. We have seen such delusions lead to all sorts of trouble, from simple refusal of medication all the way to trying to call the police to save them from the 'murderous' nursing staff, who are 'trying to poison' them. Most of the time, we can manage such situations with gentle reassurance and occasionally with drugs, and the crisis quickly abates. Sometimes, however, it can go much further than that.

Hector was 79 years old and he had suffered from breathlessness and angina for years. When no amount of medication could give him relief, he was referred to the local cardiologist to see what was causing this and if anything could be done to fix it. The cardiologist organised a few tests and found two major things wrong with Hector's heart: a severely narrowed aortic valve and a series of blockages and tight narrowings in all of his major coronary arteries. His heart muscle was weakened by all of this and he was living under the constant threat of both a heart attack and heart failure. Hector would not survive long unless something was done.

The only solution to this problem was a replacement of the aortic valve and a quadruple coronary artery bypass, quite a major undertaking in a man of his advanced age and with such a weak heart, but still better and safer than doing nothing. I had seen him in clinic a few weeks previously and we had a full and frank discussion about the options, including the risks of the operation which, as I explained to him, were substantial. He nevertheless chose to go ahead with surgery. He was admitted to hospital to have this done, and the following day I replaced his aortic valve and did the quadruple bypass. The operation had gone smoothly and, after only one night spent in the ICU, he looked very well and was transferred to the ordinary ward to continue his recovery.

In the old days, two of our surgical wards were rather unimaginatively named. The one on the ground floor was prosaically called Surgical Unit Ground Floor Ward and the one above it was called Surgical Unit Top Floor Ward. Both have since been renamed by hospital management. (Top Floor Ward is now called Mallard Ward, to show our appreciation of the duck pond and its quacking residents.) Hector left the ICU for the Top Floor Ward.

On the morning of the third day after the operation, Hector was making good progress from the heart point of view, but the nurses looking after him noted that he had mild and relatively 'friendly' paranoid delusions. With a smile on his face, he would occasionally say to one of the nurses, 'I know what you're trying to do here: you are trying to kill me, aren't you?' The nurses would reassure him that this was not the case, and if he was up and about and becoming slightly agitated about his delusions, they would coax him back to bed without difficulty. This was considered a mild case of paranoia, but easily manageable and expected to pass without incident.

Later that day, Hector again got out of bed and this time slowly made his way towards the window. The other patients in the bay watched him open the window and begin to climb through it. Across the courtyard, the nurses in another, medical, ward also saw Hector half-out of the window and hurried along the corridor to Top Floor to warn the staff. One of the nurses ran into the bay in an effort to stop him, but barely managed to reach the end of his pyjama trousers as Hector plunged 20 feet on to the concrete courtyard below.

The staff on Top Floor Ward immediately called the 'crash team' via the hospital switchboard dedicated crash-call hotline. Such a team exists in all acute hospitals and is immediately summoned if a patient has a cardiac arrest, is close to having a cardiac arrest or suffers a major life-threatening complication that demands instant, life-saving action. The crash team is composed of doctors from several specialties, but always including anaesthesia or intensive care, as well as senior nurses and technical assistants, with all the skills needed for rapid resuscitation in an emergency situation. Until a cardiac arrest or another death-heralding event occurs, the team does not really exist, and its members go about their usual business in whatever departments where they normally work. It is a team that is brought to life by death.

The hospital switchboard issued the crash call and, in scattered locations around the hospital, over half a dozen pagers in the pockets of the various members of the crash team went off simultaneously, trilling the urgent and high-pitched cardiac arrest tone, followed by the verbal message '*Cardiac arrest, surgical unit Top Floor*', repeated three times. From their various locations around the hospital, members of the crash team dropped whatever they were doing and ran at full speed, converging towards Top Floor Ward. On arrival, they were bitterly disappointed to find the patient in question was no longer there, as the incident giving rise to the crash call specifically involved his abrupt vertical transition from Top Floor Ward to the courtyard on the ground floor. They immediately ran back down the stairs to Ground Floor Ward, out of the doors to the courtyard and began to attend the unexpected casualty, who was lying on his back on the concrete. His first words to the team were, 'So, am I dead yet?'

After some early treatment and stabilisation, Hector looked well, and his heart was coping with all of this, but he was unable to move his left leg, which lay at an impossible angle not compatible with normal anatomy. X-rays showed several broken bones, including a nasty fracture of the left thigh bone (or femur) and another of the pelvis. We hastily arranged an ambulance, which took him to the Orthopaedic Department at the nearby Addenbrooke's Hospital in Cambridge.

That was enough excitement for one day, I thought, as I walked back from the accident scene to my office to deal with some of the more mundane tasks of administration, and opened my hospital email inbox.

One of the big problems of trying to run an acute hospital service in the UK is the chronic shortage of hospital beds. In fact, the UK has fewer hospital beds in relation to its population than most other developed nations. In one survey, France had seven and Germany eight beds per thousand people to the UK's paltry three. This means that hospitals in the NHS are constantly trying to discharge patients as soon as possible, and if, for whatever reason, patients do not go home as planned, admissions and elective operations often have to be cancelled due to the lack of beds.

As a result of all of this, myriad hospital management initiatives have been introduced over the years in an attempt to streamline the discharge process. Papworth was no exception, and we had, a few weeks previously, appointed a Discharge Planning Coordinator in a bid to speed up patient discharges. The first email I opened was from this coordinator and was addressed to all the surgeons. It highlighted the latest management wheeze for the speedy discharge of patients and proposed the introduction of a 'traffic light system' for discharge planning. In summary, patients were to be allocated traffic lights along the following lines:

GREEN:	ready for discharge
AMBER:	be cautious
RED:	do not discharge

The email proposed to introduce the new system as soon as possible and requested that the members of the surgical team comply with it. I was still trying to figure out how the traffic lights would help speed up discharge in practice when my email programme pinged again. David Jenkins, a fellow surgeon and friend, was providing the first surgical response to the traffic light proposal with the following missive:

> Many thanks, acknowledged, worth a try, I hope it helps
> DJ
> PS: I understand Top Floor have already developed their own very rapid discharge policy that appears extremely effective, if a little radical

It took me a few seconds to see what David was referring to, and of course I burst out laughing. Bad news travels quickly in hospitals. While I was still chortling at his wit, another email arrived from the new coordinator, who clearly had not yet heard what had happened.

They should not have done this without consultation … I'll have words with Sister.

This time I could not resist entering the fray, so I penned:

I think David is referring to defenestration.

By this time, Steve Large, another surgeon and friend, had also entered his office and seen the email exchange, so he dived in with:

I'm sorry to say that this discharge policy has been found to be burdensome by Surgical Unit Ground Floor

So I replied (I'm embarrassed to say) with:

That is the height of impudence on an issue of gravity. The low-down on all this is we must get off our high horse to deal with it and hit the ground running. I really believe this one will fly.

To which Steve wrote:

All in all a ground-breaking development in patient discharge, a leap into the unknown so to speak!

And David followed with:

One small step for patient discharge,
one giant leap for the patient in question …

I honestly have no idea for how long we could have continued this ridiculous banter, but it had to end abruptly there and then, as the next email came from the Clinical Governance office. The message informed me that the patient's son, accompanied by his lawyers, was demanding an urgent meeting with the hospital management and that my presence was required. I immediately deleted all the incriminating messages from my mailbox and advised Steve and David to do the same, but not before saving them somewhere secure for posterity. Within a few hours, the estates maintenance team were busy bolting and screwing all the windows on Top Floor Ward to prevent them from opening sufficiently to allow a person through.

A few weeks later, at my outpatient clinic on Wednesday afternoon, I was perusing the list of patients who were booked to be seen, when I immediately spotted Hector's name. I wondered what state he would be in if, indeed, he had been able to turn up to an early follow-up appointment after all these tribulations. He had. He walked in, having had his fractures fixed and having made an excellent recovery, both from heart surgery and from the defenestration injuries. He was delighted with the outcome and had dissuaded his son from taking any legal action against the hospital. He left the clinic a happy man.

At the next meeting of the European Association for

Cardiothoracic Surgery, I came across François Roques, a French surgeon and long-time collaborator in research. We went for a beer and were, as surgeons often do, exchanging patient stories. Of course, I told him this one. He looked somewhat morose to hear it, and told me that, a few years previously, a young female patient of his had done the very same thing and jumped out of a window while recovering from a heart operation. I asked if she had survived the fall. He fixed me with an unforgiving glare and said, 'Have you forgotten? My surgical ward is on the ninth floor.'

CHAPTER 5

CABG and how to avoid it

The overwhelming majority of heart operations are, fortunately, nothing like those I described above. Delivering babies and falling out of windows in association with heart operations are rare, once-in-a-career events and the overwhelming majority of operations go smoothly, with few or no complications and a boringly predictable and excellent outcome for the patient. This was not always the case. In the early days of this relatively young specialty, operations were hairy and scary things indeed. Death rates of around 10 per cent or even higher were commonplace, as were catastrophes due to malfunction of the heart-lung machine, inaccurate diagnosis and simple human error on the part of surgeons, who, despite doing their best, were flawed human beings with incomplete knowledge, feeling their way with trepidation along untrodden paths by performing pioneering procedures which had not yet been fully established, tested or made safe.

Nowadays, we know a lot more about how best to conduct an operation. We know the likely risk of it and the potential benefits. We know who will stand to gain from surgery and who will not. We also know how to care for the patient before, during and after the procedure and what complications are likely to happen. We have elaborate, well-established protocols for dealing with predictable problems and a wide armamentarium for diagnosing and dealing with unpredictable ones. In other words, most of the time, heart surgery is really routine stuff, and much of it deals with the consequences of furred-up coronary arteries. But why do these arteries fur up in the first place? Is there anything you should do to avoid the potentially disastrous consequences of coronary artery disease?

Many people care a lot about their health, and quite rightly so. For those who do, a large section of their concern is concentrated on the heart, and specifically on the avoidance of coronary artery disease. This is hardly surprising: coronary artery disease is common, responsible for much premature death, and is the Number One killer in the Western world. What is it? And why and how do we get it?

The job of the heart is to pump oxygen-rich blood to every organ in the body. Without that blood, the organs will die. The blood is delivered to the organs through the aorta and its branches, the arteries (*artery* means a blood vessel coming directly or indirectly *from* the direction of the heart, and *vein* means a blood vessel heading directly or indirectly *towards* the heart). Every bit of the body needs an artery to feed it, and the network of arteries supplying the human body is truly awesome. If an important artery feeding an organ or a limb is completely blocked, that organ or limb will die. If that happens in an artery feeding part of the brain, the result is a stroke, which kills a bit of the brain. Elsewhere in the body, similarly unpleasant things can happen when an artery is blocked, like dead gut, dead leg and so forth.

Like every organ in the body, the heart needs a blood supply. In fact, on a weight-for-weight basis, it is quite greedy and needs a lot: unlike other muscles in the body, the heart beats constantly, whether you are running a marathon, having sex or just sleeping. While you rest, your heart does not. The heart gets its blood supply from two arteries called the left and right coronary arteries. They are so named because it looks as though they encircle the heart like a crown, and are the very first two branches of the aorta, just a few millimetres downstream of the aortic valve. In fact the left coronary artery is really short: one or two centimetres at most, and then it divides into its two major branches: the left anterior descending artery (or LAD for short) and the circumflex, which is the other branch. Most doctors and informed patients therefore think in terms of three coronary arteries: LAD and circumflex (two branches of the left) and the right coronary artery. Hence we have triple vessel disease, triple bypass and so forth.

The trouble with the coronary arteries is that, as time goes by, they fur up. That's not all that surprising, since all arteries do the same with increasing age. The relentless pulsating pressure within them and the constant battering from the turbulent blood flow affects their walls, and plaque builds up within the artery. In fact, no human will reach a ripe old age without at least some sign of furring in the arteries. The particular trouble with furred up coronaries is twofold: first they are really small. An LAD, circumflex or right coronary which measures more than two to three millimetres across is a whopper by my standards. So it does not take that much furring up to block one of these arteries or at least to reduce the blood flow down to a level that causes mischief. The second is that they are important because they supply the heart muscle and if one becomes blocked it leads to death of part of the heart, otherwise known as 'myocardial infarction' or heart attack. If they are narrowed sufficiently to hamper the blood flow, the heart complains about the lack of oxygen by hurting: a dull ache appears across the chest, especially on exercise. This is angina and its usual cause is coronary artery disease. Is it possible to avoid it?

The link between this disease and lifestyle choices is firmly established in the mind of the public and the doctors. To that end, we see health-conscious people going to the gym, jogging, cycling, staying generally fit, avoiding cigarettes, watching their alcohol intake and carefully controlling what and how much they eat. They are also hugely interested in any research titbits that may hold out the promise of a healthy heart and, by implication, a longer life. The mass media ruthlessly exploit this interest, almost as ruthlessly as they exploit fear of cancer. Thus every snippet of research that shows this or the other lifestyle measure might reduce your risk of coronary disease is seized on by the mass media, fed by researchers who also seem obsessed with this particular topic. Just about every minute aspect of diet and lifestyle has been analysed in relation to its likely impact on your coronary arteries. This has resulted in a veritable avalanche of scientific information, which, in turn, has generated a profuse amount of advice, available from every conceivable source: not just from the GP surgery, hospitals, researchers, royal colleges and medical journals, but also from dieticians, lifestyle advisers, radio and television programmes, glossy magazines and most of our daily newspapers. Some of this advice is sound, some of it is not and some is in direct conflict with other advice and with the evidence. So how on earth do you decide on a healthy-heart lifestyle?

Before asking the heart question, let us ask a more general but highly pertinent question about the way we approach this issue: how do we, as ordinary folk, make decisions about the risks to our health? The answer is that most of us do it badly. We are not sufficiently well informed about the real science behind the risk of heart disease. Sometimes we have good information, sometimes we have bad or wrong information and sometimes we simply do not have the information. The imperfect state of our knowledge is further distorted by two major factors: sensationalist and highly imbalanced reporting in media scare stories and a 'health lobby' that seems almost religiously puritanical in the advice that it dishes out. Finally, most of us are not sufficiently educated in understanding and interpreting risk, and fall prey to many fallacies of 'heuristics' — the way the human brain evolved to learn and come to conclusions. When all of that toxic mix of misinformation and misjudgement is taken into consideration, it is a miracle that we make any sensible decisions about our health at all!

Let me start by telling you about an evil, cancerous tumour of the heart. It is horribly malignant and goes by the name of cardiac fibrosarcoma. It is a most unpleasant disease. There is no cure for it: this cancer is almost universally fatal and, despite using all available cancer treatments, such as surgery, chemotherapy or radiotherapy, the time between diagnosis and death is, on average, only about 11 months. What's more, death from cardiac fibrosarcoma is a gruesome and painful process. This is definitely one disease to avoid. Now, let us open a newspaper, like the *Daily Mail*, which revels in disseminating the latest breaking news in medical research, especially where cancer is involved. Today's issue, hot off the press, has a headline on the latest top scare story. It says that a recently published scientific article has just reported the results of a study which shows that drinking just one cup of coffee a day increases your risk of cardiac fibrosarcoma by 92 per cent.

Will you now stop drinking coffee? Of course you should! You care about your heart and your health, don't you? Who the hell wants to end up dying from cardiac fibrosarcoma for the sake of a cup of coffee? On the face of it, throwing away your coffee-maker would appear to be a rational decision, but with the level of evidence that we have so far, it most certainly is not. Let me explain.

The first question has to be this: how common is cardiac fibrosarcoma? The answer is: not at all. Cardiac fibrosarcoma is one of the rarest tumours that arise from the heart, and accounts for fewer than 3 per cent of all primary heart tumours. Tumours which arise from the heart are themselves also vanishingly rare, being found in around only 0.001 per cent of the population, meaning that the overall risk for one person to have a diagnosis of cardiac fibrosarcoma is 3 per cent of 1 in 100,000 or around 1 in 3 million. The newspaper article reports what sounds like a terrifyingly massive increase of 92 per cent in this risk, but that takes it from a paltry 1 in 3 million to a slightly less paltry risk of just under 2 in 3 million.

Is that enough to give up coffee? It is definitely not. Make mine a double espresso, please, and, by the way, there is, to my knowledge, no link whatsoever between coffee and cardiac fibrosarcoma, and nor is there a reason why there should be such a link, but the mass media are full of similar scare stories. Of course, responsible reporting of such research is possible: an honest, non-sensationalist headline could say 'Coffee might slightly increase the tiny risk of a very rare disease, but that risk is still tiny', and that would be absolutely true, but it would not do much good for newspaper sales and shows why I would be an utter failure as a headline writer. The next time you are given information that something has been found to increase your risk of some dreaded lurgy by umpteen per cent, your first question should be: umpteen per cent of what, exactly?

Here is another lurid (but fictional) health news story:

DIET COKE® MAKES YOU FAT!

Scientists have proved that the consumption of low-calorie fizzy drinks is associated with obesity. In a major scientific study carried out at Dodgy University, researchers stopped people who walked past their office holding a fizzy drink and asked them to step on scales in order to measure their weight. They found that those who had low-calorie drinks weighed, on average, 19 kilograms (or three stone) more than those who had sugary drinks. Plans to introduce a sugar tax by our nanny state are sure to be derailed by this new finding. The Minister of Health was not available for comment.

For once, I do not blame the Minister of Health. What is blindingly obvious about the way this so-called study is reported in the newspaper is the false assumption that association means causation. Or, in other words, if something A happens when something B is around, then B must have caused A. That is not necessarily the case. Of course, B may have caused A, or A may have caused B, or both A and B may have been caused by something else, C. Thus there are three possible explanations for the 'link' between obesity and low-sugar drinks. The first is that low-sugar drinks cause obesity, as the newspaper automatically assumed. The second is that obesity causes low-sugar drinks: you get fat, start to worry about your bulk, and start to buy diet drinks. This is almost certainly the explanation here, but it still needs proper scientific confirmation. The third is that obesity and the consumption of low-sugar drinks are both caused by a third factor (cultural or genetic, perhaps).

Now you might think that I am being facetious in exploiting this fictional example, but a kind of thinking very similar to this has had a large influence on the behaviour of the health lobby, such as the supposed link between lower levels of heart disease and, for example, eating a Mediterranean diet or moderate consumption of red wine. They could be related by causation, but not necessarily.

Finally, the health lobby and the Government must also shoulder some responsibility for the misleading advice that has proliferated so extensively in the last few decades. The NHS Choices website has a section entitled 'Ten Tips for a Healthy Heart'. It quite rightly advises you to stop smoking, exercise more and not get fat. It also advises you to reduce your salt intake. This is, in my opinion, manifestly wrong advice.

The evidence for a link between smoking and coronary disease is overwhelming: overall, a smoker is twice as likely to have a heart attack as a non-smoker, but coronary disease is quite prevalent in old age. If we look only at people under the age of 50, a smoker is five times (that's a whopping 500 per cent) more likely to have a heart attack than a non-smoker; and, remember, this is not a rare condition like fibrosarcoma: heart attacks and coronary disease are very, very common. We also understand many of the mechanisms by which smoking causes heart disease: the inhaled chemicals directly damage the lining of the coronary arteries. In fact, they damage the lining of all arteries, and their effect on other arteries, say, those to your brain or to your legs, is even worse than on the coronary arteries. It is true that nobody has yet done a randomised controlled trial in which people were randomly made into smokers to see if they would die earlier from heart disease, because with the level of evidence we already have, that would be unethical. However, many interventions designed to stop people smoking have been seen to reduce the level of coronary heart disease, and the dramatic drop in this disease that we have witnessed in the last two decades is rightly believed to be in large part due to the decline in smoking in the general population. When you then consider that, on top of coronary heart disease, smoking has been proven to cause lung cancer and a host of other unpleasant cancers, the NHS Choices website is fully justified in advising us to ditch the cigarettes. But is the website justified in its recommendation to reduce salt intake?

There, the evidence is not overwhelming. In fact, you could almost say that there is hardly any good evidence at all to back this advice. In medical science, the best evidence is obtained from randomised, controlled trials. That means that a number of people are divided at random into two groups: you give one group treatment A, which is the one you are interested in studying, and the other group treatment B, which could be a placebo or no treatment. Then you watch and measure the outcome you are interested in, such as death or heart attack.

Even better than a randomised, controlled trial is a 'meta-analysis' of many randomised, controlled trials: in a meta-analysis all the good-quality trials that have been done are pooled together in order to reach a conclusion. The Cochrane Collaboration is a non-profit organisation largely funded by the National Institute of Health Research. It is highly respected as one of the world's best in the field of performing meta-analyses, with the aim of distilling the best scientific evidence out there, so that the soundest conclusions can be reached about important medical topics. In 2014 there was a published Cochrane review on the effect of reducing salt intake on survival, cardiac events and blood pressure. To do this, they analysed seven randomised clinical trials in which salt reduction was treatment A and no salt reduction was treatment B (the control). The following is from the abstract of their report, in which each scientific statement on the left is accompanied by my explanation, in plain English, on the right:

SCIENCE TALK	PLAIN ENGLISH
Seven studies were identified: three in normotensives, two in hypertensives, one in a mixed population of normo- and hypertensives and one in heart failure.	We found seven trials. Some looked at people with high blood pressure, some at people with normal blood pressure and some at people with heart failure.
Salt reduction was associated with reductions in urinary salt excretion of between 27 and 39 mmol/24 h and reductions in systolic BP between 1 and 4 mm Hg.	People told to cut down on salt actually did so, because they passed less salt in their pee when we measured it. Also their blood pressure came down a tiny bit.
Relative risks (RRs) for all-cause mortality in normotensives (longest follow-up - RR: 0.90, 95 per cent confidence interval (CI): 0.58–1.40, 79 deaths) and hypertensives (longest follow-up RR 0.96, 0.83–1.11, 565 deaths) showed no strong evidence of any effect of salt reduction on CVD morbidity in people with normal BP (longest follow-up: RR 0.71, 0.42–1.20, 200 events) and raised BP at baseline (end of trial: RR 0.84, 0.57–1.23, 93 events) also showed no strong evidence of benefit. Salt restriction increased the risk of all-cause mortality in those with heart failure (end of trial RR 2.59, 1.04–6.44, 21 deaths).	Reducing salt made no difference to survival in people with high blood pressure. It also made no difference to survival in people with normal blood pressure. It did make a difference in people with heart failure: reducing their salt intake made their survival worse.
We found no information on participant's [sic] health-related quality of life.	Quality of life was not studied.

When you read this, you may wonder how on earth the Government and the health lobbies can possibly pursue salt in the diet with such messianic zeal. In fact, what is even more amazing is that the Cochrane review, having found that there is absolutely not a shred of evidence to support telling people to reduce their salt intake, concludes:

> Our findings do not support individual dietary advice as a means of restricting salt intake.

The conclusion above is located in the scientific bit of the article. In the very next paragraph, there is a summary of the findings in plain language and this is what it states:

> The findings of our review do not mean that advising people to reduce salt should be stopped.

Really? And why not? And how come the conclusion that is aimed at scientists appears diametrically opposite to the plain-language conclusion?

Salt may of course still be harmful, or it may not be. It should, however, be considered innocent until it is proven guilty, because there is a risk of diluting evidence-based advice — such as 'Don't smoke' and 'Don't get fat' — by broadcasting wholly unimportant and unproven messages like the one about avoiding dietary salt.

This is from the official health lobby. The unofficial sources of advice in newspapers, magazines and other media also tell us to avoid crisps and butter and red meat, and to eat quinoa and organic pasta and drink cranberry juice smoothies and detox tea and fat-free soya milk and other suchlike fashionable fare. I truly believe that — in the absence of any good evidence — we doctors and health-care professionals should strive to keep our advice firmly on the shelf and let people live the way they want to live. Firstly, one should not be haranguing people to change their lifestyle without good reason. More importantly, giving ten lifestyle instructions of which only one is valid and the other nine are of dubious value is one sure-fire way of putting people off following the instruction that actually matters. Too much advice can be self-defeating.

I have chosen to focus on the dietary advice on salt for a simple reason: it is the one bit of advice that is most commonly given and with the least supportive evidence, and yet it continues to be dished out to the population freely, frequently and quite ferociously. Salt is not the only victim to be convicted by such a miscarriage of justice: other convicts may also be innocent or, at the very least, there may be serious doubts about their conviction. These include butter, dairy products, red meat and, yes, even cholesterol.

So how do we — in the face of all of this conflicting evidence — make decisions about lifestyle with a view to protecting ourselves from coronary artery disease? I think the answer is quite simple, but the view I hold will not necessarily be endorsed by the health lobby. Here it is anyway:

The first thing we all have to realise is that many of the risk factors for coronary artery disease are utterly beyond our control, and there is no lifestyle modification on earth that would make a jot of difference to them. They are:

- Genetic make-up (who your parents are and the prevalence of coronary problems in your immediate family)
- Age (the older you are, the more you are likely to have coronary disease)
- Sex (males fare a little worse off than females in the coronary stakes)
- Bad luck

For the above group of risk factors, we just have to accept that there is not a lot we can do.

Secondly, there are risk factors that we can definitely do something about, and these are proven beyond doubt:

- Smoking tobacco is very, very bad, and that is a fact
- Being fat is very bad, and that is a fact

For this pair of risk factors, the answer is simple: just don't do it. Do not smoke tobacco and do not allow yourself to become fat. Put the cigarette out and, having eaten enough, put the fork down: quite apart from the coronary disease, a lot of other bad things can happen to you if you do not.

It is true that different people have different mechanisms for detecting that they have had enough food, and that upbringing, culture and societal pressures all play a part, but the bottom line is that if a person consumes more calories than he or she burns as fuel, the extra food will be deposited as fat. This means that as we put on weight, the solutions are to eat less, exercise more or, preferably, both.

There is a further complicating factor as we age. Many people will have noticed that while they could eat large amounts of food in their late teens and twenties without putting on an ounce of extra weight, the same does not apply after the age of 30, and from about that point onwards extra food seems quickly to translate into excess weight. There is a simple scientific explanation for this phenomenon: from the age of around 30 onwards, we naturally begin to lose muscle mass. The process even has a name: age-related sarcopenia. Losing muscle is a bad thing for many reasons. It affects strength and mobility and of course makes exercise harder and less desirable, and that leads to inactivity and more muscle loss. However, loss of muscle mass also has a direct effect in that it can make us put on weight. The reason for this is simple.

Muscle is a relatively live and active tissue with lots of blood vessels and therefore burns many calories. Fat is not, being mostly a storage depot for yellow globules with relatively little metabolic activity and hardly any blood vessels. This means that if you have two people weighing 80 kilograms each, the one with more muscle will burn more calories when merely standing still or sleeping than the one with less muscle. The less muscled person will have to reduce food intake much more to maintain a healthy weight than the one with more muscle. In other words, if you want to carry on eating the same amount of food that you did in your teens, you will need to exercise in a way that preserves your muscle mass. That means some resistance training and exercises aimed at preserving the large 'core' or antigravity muscles, such as Pilates. We have a choice: let muscle loss continue, but eat a lot less, or fight it with resistance training and eat more. I chose the second option, because I really like food.

Thirdly, there are proven risk factors that we can modify, even if we cannot get rid of them completely:

- Having high blood pressure is bad. Sometimes just exercise can reduce your blood pressure, but if it does not, take the tablets!
- Being diabetic is bad. If it is type 1 diabetes, you can't do much about it, other than making sure you keep your blood sugar well controlled. If it is type 2, you may be able to reduce its severity or even get rid of it altogether by shedding some weight.
- Having high cholesterol in the blood *may* be bad, but this is a surprisingly complex area of science. Here is what we know: the plaques that fur up our arteries hold quite a lot of cholesterol and cholesterol-containing compounds. On the face of it, eating less cholesterol might therefore help reduce the burden of cholesterol-laden plaques in our arteries. However, it is not conclusively proven yet that high cholesterol in the blood leads to coronary artery disease in everybody. In other words, we are not sure that cholesterol in *plaques*

> always comes from cholesterol in *blood*. Furthermore, it is also not conclusively proven that cholesterol in the *diet* is linked with cholesterol in the *blood* (most of the cholesterol in the body is actually made by your liver and for a good reason: your tissues need it, especially your brain). However, if you want to modify your blood cholesterol it is a hell of a lot easier, quicker and more effective to achieve that by taking a statin tablet than by any diet modification, no matter how draconian. This means that if you have high cholesterol and, especially, if you are in a high-risk group (such as having either already been shown to have coronary disease or carrying many of the risk factors above which you can do nothing about), then it makes sense to take a statin (and then eat whatever you like, as long as you don't get fat). If you belong to one of these groups and you can tolerate statins (and most people do so with relatively few side effects), then you should probably take them. They will not harm you and may well do you some good. I happen to be in a high-risk group (see Chapter 16) and I do take them.

Finally, we come to the difficult area: those things which are not proven or where any proof is still controversial. Here it becomes a personal issue. Many considerations have to be taken into account, not least being our varied and highly individual approach to risk. From my point of view, we choose our lifestyle because (I presume) it is the kind of life that we want and enjoy. We only have one life, and of course we want it to last, but we also want it to be a happy and enjoyable life while it lasts. How much we are prepared to sacrifice some of the enjoyment for the (uncertain) prospect of lengthening it just a little bit is a very personal decision. We all know that skiing, riding a motorbike and sailing the oceans can be dangerous, and they are all much more dangerous than sitting on a sofa at home, but try to persuade an avid skier, biker or sailor that they should abandon their activities for the unexciting prospect of the sofa! My own take on this issue is that I will categorically refuse to modify my own lifestyle for any of the following:

- Anything which is still not proven to cause harm
- Any risk of getting something which is vanishingly rare, no matter how horrible it is
- Any risk of getting something which is common, but which is not all that horrible and which I figure I can live with quite comfortably

Within the first category above are all the so-called 'risky' diet choices that the newspapers, magazines and health lobbyists never cease to warn us about: red meat, butter, cheese, crisps and, yes, of course, salt. I eat all of these to my heart's content and you can probably do the same as long as you do not become fat.

Putting it another way, it is not so much what you eat, but how much!

CHAPTER 6

An easy cabbage

The routine operation that we do most commonly is coronary artery bypass grafting, or CABG for short. We surgeons refer to it as a 'cabbage', a name that most people of the politically correct persuasion do not really like. I am not wholly wedded to the politically correct movement, but can recognise in this instance that referring to the 'cabbage in bed 7' within earshot of the patient's family is probably suboptimal.

CABG is, above all, an operation to relieve angina due to narrowings and blockages in the coronary arteries. If you are an otherwise fit patient with angina and need a CABG, you will come into hospital the day before or even on the morning of your operation. You will have a general anaesthetic and a number of tubes put in for the purposes of monitoring and delivering drugs. You will then be wheeled into the operating room. There, a surgeon will open your chest with a saw, cutting the entire breastbone lengthwise. From within, the surgeon will take down the left internal mammary artery (the best bypass tube possible for the coronary artery on the front of the heart). At the same time, another surgeon will take out a piece of vein from your leg. How long that piece of vein is will depend on how many additional bypass grafts you need: for one bypass, a piece the length of a ballpoint pen will do. For three or more, the whole vein from ankle to groin may need to be used. Once these conduits (mammary artery and leg vein) have been 'harvested', the surgeon will put you on the heart-lung machine, clamp your aorta and give the potassium solution to stop and preserve your heart.

First the bits of vein and last the mammary artery will be joined to your own coronary arteries beyond the blockages using a very fine stitch of polypropylene. Each join will take 10 to 15 minutes. The mammary artery bypass graft will then work straight away, as its origin is still attached to the arterial network. The bits of vein will need to be plumbed into the arterial circulation, and they will be stitched to the aorta, the nearest artery accessible in the operating field. Now that the bypasses are working, the heart-lung machine is taken away and the chest is closed over some drains, which will let out any blood that is lost after the operation is finished, and also usefully serve to monitor just how much blood loss there is (if it's excessive, a return to the operating room may be needed).

Most CABG surgery is routine, predictable and relatively safe. That does not mean, however, that it is an operation that is free from risk, and, like so many other medical interventions, it can occasionally be extremely challenging. In fact, most of the headline-grabbing operations in heart surgery — such as heart transplantation, heart-lung transplantation and implanting artificial hearts — tend to be technically not too demanding. They take time and care, of course, but they consist of operating on, handling and stitching relatively big structures, so that the technical ability required is well within the reach of most surgeons. By contrast, a coronary artery may be only one millimetre in diameter, and, if it is heavily diseased, buried deeply within the heart muscle, calcified or inaccessible to the eye or the hand, bypassing it becomes a technical challenge that is fraught with danger. Moreover, it only takes a single stitch that is a mere quarter of a millimetre too deep, or one that erroneously and accidentally picks up the opposite wall of a coronary artery, or a dog-ear in an imperfectly completed join for the coronary artery to end up blocked rather than bypassed. A blocked coronary artery means a heart attack and a heart attack on top of a heart operation is often too much for the heart to bear, so death usually follows. Fiddly and small coronary arteries make this more likely, and so do those with diffuse disease and calcium within their walls. Whenever I am asked which, of the many different operations in our surgical repertoire is the one that I find most technically demanding of all, my reply is invariably 'A difficult CABG', and I think many of my colleagues would agree with me.

With the operation over, you will be taken to the intensive

care unit for a period of close observation, where your pulse rate, blood pressure, ECG, temperature and urine output will be intensely monitored. The nurse looking after you will also do regular blood tests to make sure your lungs and kidneys are working properly and to measure sugar and metabolism, all of which give an idea of how happy the heart is with the outcome and how happy your body is with the function of your heart. Two to three hours after arriving in the ICU, if all is well, and if there is no excess bleeding, the nurse will switch off your anaesthetic drugs and you will wake up soon afterwards. You will notice, through the haze of the receding oblivion of anaesthesia, that there is a tube at the back of your throat delivering breaths from the ventilator. By the time you are awake enough for this tube to begin to bother you, you are ready to breathe on your own and the tube is taken away. Because your belly has not been touched, the 'nil-by-mouth' rule does not apply, and you can have a cup of tea when you feel like it. The following morning, you will sit in a chair, have breakfast and be taken back from the ICU to an ordinary ward bed. The day after that you will get up and walk around your bed. On Day 3 you will venture a little further with the help of the physiotherapist. On Day 4 you can walk the length of the corridor and by Day 5 you can go up and down a flight of stairs and you are ready to go home. When you get there, you will feel tired and not up to much, but this does not last long, and when you finally decide to exert yourself physically, you will be delighted to see that the angina, which had previously limited your lifestyle, has now completely gone.

This is what most heart operations are like. CABG accounts for around half of the heart surgery done worldwide, and the majority of patients, especially if they are relatively young (that's under 75 in my book) and otherwise fit, will sail through the procedure and its aftermath with not an ounce of drama.

The standard CABG patient is male, middle-aged or older, a little overweight and may have diabetes, high blood pressure and high cholesterol. He may have been a smoker and there will have been a history of coronary disease in the family. Liam Hughes, a close friend who at the time was a cardiologist in Norwich, referred me a patient for CABG who was as far removed as one could be from the standard CABG patient. Her name was Emma Chapman. She was young: only 35 years old, slim, with no diabetes or high blood pressure, no high cholesterol and no family history of heart disease. Her only risk factor was that she had been a smoker, hardly sufficient to cause coronary heart disease in one so young, and yet she had angina and quite severe breathlessness whenever she exerted herself and was mightily fed up with that state of affairs. At first, Liam thought she could not possibly have coronary disease, but her symptoms were so convincing that he eventually relented and carried out an angiogram: he put a catheter in the artery in her groin, threaded it up the aorta to the origin of the coronary arteries and injected some dye to see what they looked like. There was no doubt about it: one of her coronary arteries was completely blocked. It was the artery on the front of the heart, the left anterior descending artery or LAD.

The LAD is one of the most important arteries in the body. It supplies the front of the heart, including a big chunk of the left ventricle, so that when it is blocked, a sizeable piece of heart muscle may die and the heart attack that follows is often a big one. Cardiologists used to call the LAD the 'widow-maker'. Emma's LAD was blocked completely, but her heart was still in good shape, so that she was lucky that whenever that blockage had happened, there was not much muscle loss, and her heart had continued to function almost normally. Nevertheless, one part of her heart muscle was being deprived of blood and oxygen, and every time she tried to do anything physical she had to stop because of angina and breathlessness. She wanted something done.

From the point of view of survival, a CABG would not help her, because she was not at any risk: the LAD was already completely blocked and therefore simply could not get any worse. She had already survived the blockage, and her other arteries were in reasonably good shape. Since there was nothing in her coronary arteries at that time that would pose a risk to life, an operation was being contemplated only to get rid of her angina and improve the *quality* of her life. As she was fit and young and with no other risk factors, her risk of death from an operation would be less than 1 in 100, and it was on that basis that I saw her in my clinic. We talked about her having the operation and I explained the risk that she would take. She thought that a risk of less than 1 per cent was most definitely one that was worth taking to get rid of her angina, and we agreed to proceed. Having made that decision, the next one was about the surgical approach, or where to make the incision.

Most heart operations require full access to the entire heart and are therefore done through a long up-and-down incision in the middle of the chest, from the top to the bottom of the breastbone. Occasionally, it is possible to do such operations through smaller incisions, and this was such an occasion. The LAD is towards the front of the heart. The left internal mammary artery, which will be used for the bypass, is behind and to the left of the breastbone. In fact, the two arteries are no more than 2 to 4 inches away from each other. It is therefore just possible to do the bypass through a much smaller incision under the left breast: go between ribs, separate the mammary artery from the chest wall, open the pericardium (the bag around the heart) and if the LAD is visible and easily accessible, join the two together while the heart is beating, without using a heart-lung machine. The great advantage of this 'minimally invasive' approach is that the body hardly notices that anything has happened, and the recovery is very quick indeed. The other advantage is that the incision is almost invisible and well hidden by a bra or bikini top.

The disadvantage, of course, is that it is technically more challenging, and that topological reasons can make it impossible, such as an LAD that is too far away or too well hidden within the heart tissue. That sort of thing can only be ascertained by making the small incision and having a look. If it turns out to be impossible or dangerous to proceed, the only option is to close the small incision and make the big one, so that the patient would end up with two incisions instead of one. In my practice, I have had to convert roughly between 1 in 10 and 1 in 20 such operations to the big incision, and I explained all that to Emma and gave her the choice. She opted for the small incision. Since she was slim, and her angiogram had shown the LAD clearly to be not too far away from the midline, I was quietly confident that the small incision would work. I put her name on the waiting list.

Emma came into hospital on a Monday morning for the operation to be done in the afternoon, the second (and last) operation of the day. Jon Mackay, the consultant anaesthetist, was putting her to sleep and I was checking on the first patient I operated on that day, who was recovering in the ICU, when Jon paged me. I went to the anaesthetic room and found him quite agitated. As soon as he had put her to sleep, Emma had started to have heart rhythm disturbances, with the heart beating irregularly and very fast.

'I'm really not happy with you doing this through a minimally invasive approach,' he said. 'Her heart is all over the place and she is unstable. Do the full incision so that we have more control.'

I considered what Jon had said, and it simply did not make sense: here was a woman with one completely blocked coronary artery and all the others were healthy. The blocked one couldn't get worse and the others were fine. Why on earth should she become unstable? And why would opening her chest fully help in any way? I disagreed with Jon, suggested that he use drugs to control the heart rhythm, which, I surmised, must be a strange reaction to the anaesthetic, and that we proceed as planned. Jon was not overly happy with this approach, but eventually I persuaded him. He gave some drugs to slow down the heart rate, and we proceeded.

I made the small incision under the left breast and dissected out the mammary artery. I opened the pericardium and the LAD was there, exactly where I wanted it to be, but it was very small, barely one millimetre in diameter. I remember thinking 'Really? Was all this angina and breathlessness a result of a blockage in *this* tiny blood vessel?' I pushed that thought out of my mind and proceeded to join the two arteries together, while the heart went on beating happily and regularly. Jon Mackay had calmed down, everything went well and, less than two hours after starting, the operation was finished and the wound was closed. I was happy with the result, forgot about the heart rhythm disturbance and the smallness of the artery, and Emma was taken to the ICU.

That evening I was to look after our two daughters, while my wife attended a teachers' meeting at the school with our two older sons. I reached home just in time for the school expedition to make it to the meeting, and settled with the girls in the living room, chatting, playing games and watching TV. An hour later the phone rang. It was the registrar on call that night for the ICU, informing me that 'Your young patient Emma Chapman' had, with no warning, suffered a cardiac arrest. She had gone into ventricular fibrillation, a heart rhythm in which the heart stops beating altogether and its individual muscle fibres just wriggle at random with no heart function whatsoever. The staff had immediately given her an electric shock, and she had snapped out of it instantly. She was fine now and everything was looking good, but the registrar felt that 'I would want to know.' My own heart sank.

I simply could not understand the reasons for this sudden instability. I had a patient with a blocked artery and nothing else. That artery now had a bypass graft. What on earth could have happened — two hours after a smooth and uncomplicated operation — to make her do this? Even if I had botched the procedure and my bypass graft was entirely useless or had suddenly clotted off, she should not be in a worse state than she was before it was done. This just didn't make sense.

It was not easy, trying to juggle all these thoughts while still trying to be a devoted and attentive parent. Over the next half an hour or so, I somehow managed to convince myself that this was a freak event and that Emma and her heart would behave as expected from now on. I tried to focus my attention on my daughters in what should have been a rare and valuable evening when we would actually have time together.

The phone rang again.

This time it was not the registrar calling, but one of the ICU nurses, and, because of that, I knew instantly that some disaster must have taken place. There is no worse omen than a call from the ICU nurse: it means that, whatever the problem that the registrar was dealing with, it was too serious to permit leaving the patient to make a simple phone call. The problem must be of catastrophic proportions. It was.

'She's had another cardiac arrest, ventricular fibrillation again, and we can't "shock" her back. We're doing external cardiac massage, and we need your help.'

The next few minutes were a blur. I made a barely coherent call to a friendly neighbour and asked her to babysit at very short notice, jumped into the car for a furiously fast drive the 12 miles or so to the hospital. Throughout the journey I had the car phone connected to the ICU, so I could keep track of what was happening. External cardiac massage was continuing apace and an operating theatre was being prepared at my request. For the next 15 minutes or so, there was nothing constructive I could do. I just had to get to the hospital quickly, and during that time my mind was experiencing an avalanche of thoughts and emotions.

Many years later, I studied the reactions that we surgeons have in response to unexpected catastrophes, and the pattern of behaviour that we all go through on first realising that something has gone terribly wrong. It is not rocket science to recognise the pattern as virtually identical to the well-known five stages of grief, described by the eminent Swiss psychiatrist Elisabeth Kübler-Ross. Her classification, of course, referred primarily to the psychological human reaction to death and dying, but it is easily applicable to any other form of unmitigated disaster. So here are the five stages, applied to this particular surgical disaster:

1. **Denial (and isolation):** My first thoughts were: 'This cannot be happening! I don't believe it. Maybe it's all a mistake. I'll get there and all will be fine. Perhaps it's a prank … ' and so on. I felt vindicated when I found out that later versions of the five stages have added 'isolation' to the denial stage, and there is no doubting the overwhelming feeling of loneliness that engulfs a surgeon facing such a situation. One way or another, as we surgeons caused the crisis by operating, the buck stops nowhere else but at our door, and, despite the valuable support and help we receive from the many other professionals involved in surgery, when it comes to the feeling of ultimate responsibility, we are alone. The isolation that such situations engender is overwhelming.
2. **Anger:** Having accepted that it is happening, I became frustrated and angry: 'Why me? It's not fair!'; 'How can this happen to me?'; 'Who is to blame?'; 'Why would this happen?'
3. **Bargaining:** 'If only she pulls through, I promise never to do a minimally invasive procedure again, and I will always listen to the advice of the anaesthetist, and I will be a better person, lead

a better life, be courteous to people and kind to animals and so forth … '

4. **Depression:** 'I'm so fed up with this, why bother with anything?'; 'She's going to die, so what's the point? Why go on? Why do I even bother? There must be better ways of earning a living than persevering in this awful profession.'
5. **Acceptance:** 'It's going to be OK. I can't change what has already happened, so I may as well prepare for the next stage. Nothing is impossible.'

A cursory look at these five stages will show that the first four are utterly unhelpful in getting out of a fix. In fact, they are little short of being manifestations of temporary madness. But they simply must be traversed before acceptance is reached and some form of sensible and rational action to remedy the situation can at last begin. I have learnt and, at seminars and conferences, taught colleagues and trainees all the essential tricks to ensure that they can speed through the first indulgent yet unproductive four stages as quickly as humanly possible, so that they can again become an effective force in saving their patients' lives. That evening, I think I had already crossed the first four stages by the time I arrived at the hospital. It was half-past nine at night. I parked the car appallingly badly, as close to the theatre block as possible, ran into the changing rooms, put on scrubs and leapt up the stairs to the operating theatre just as the patient was being wheeled into it on her bed, escorted by half a dozen people, drip stands, ventilator and the other paraphernalia of ICU.

Maura, one of the ICU nurses, was kneeling on the bed astride Emma's chest and administering external cardiac massage during the hurried transfer. We moved Emma's lifeless body on to the operating table, threw a gallipot-full of skin preparation solution on her chest (and on Maura's hands), draped her in a hurry and proceeded to open the chest. No bikini-covered minimally invasive incision this time. We needed to get her on to the heart-lung machine as quickly as possible to ensure her brain stayed alive, while we tried to fix whatever it was that had gone wrong with her heart. So I cut from the top of the chest to just above the navel, sawed through the breastbone and exposed the heart. External heart massage was then replaced by internal massage, with my right hand around the heart squeezing it to produce a heartbeat, and using my left hand to insert the two big pipes needed to connect to the heart-lung machine. All this was done in a few minutes and I was relieved to be able to say to the perfusionist: 'OK, go on, bypass.' The machine started. The blue blood went from the right atrium out of the patient, and the pink blood came back into the aorta. 'One litre … two … three litres … we have full flow,' said the perfusionist, and I was thankful that finally I was able to stop squeezing the heart. My right hand was sore and needed a rest.

I now looked at the heart properly for the first time. The bypass graft was still attached, which was something of a miracle in view of all the external cardiac massage, the frantic rate at which we opened the chest and the subsequent internal massage. The heart was in ventricular fibrillation: each of its muscle fibres was wriggling independently and the heart looked like a bag of worms, plenty of movement but no useful synchronised pumping. The first thing to do was to 'shock' the heart out of this useless action, and I requested the defibrillator paddles. I placed one paddle in front of the heart and one behind.

'Charging to 10 joules,' said Amo Oduro, the consultant anaesthetist, and then: 'Charged.'

I pushed the button on the paddle. The heart jumped, settled and started beating normally. Totally normally. With every part of it contracting lustily, with no areas of weakness, nothing to indicate shortage of oxygen, a normal electrocardiogram on the monitor and, in short, everything pointing to a healthy and happy heart. We all looked at one another in amazement. 'What the fuck was all that about, then?' was the unspoken thought of everyone in the operating theatre at that moment.

The heart had stopped, in total, for just over half an hour, and had been subjected to vigorous massage, externally and internally, to keep some sort of blood flow for the patient to stay alive. It would therefore be a good idea to give the heart a bit of a rest on the heart-lung machine before demanding that it take up its pumping duties in supporting the circulation once more, but I wanted to check that the bypass graft was still functional. We reduced the flow in the heart-lung machine temporarily and the heart readily took over the circulation. I was easily able to feel a healthy and vigorous pulse in the bypass graft. Thus reassured, we went back to 'full flow' on the machine and I resolved to give the heart 30 minutes of rest before coming off the machine. With that resolution, there was the welcome opportunity to descrub, step out of the operating theatre and have a coffee. I took it.

Half an hour later, refreshed and relieved, I scrubbed again. Amo switched on the lungs and the patient came off the heart-lung machine without the slightest difficulty, with her heart willingly pumping to support the circulation without the use of any stimulating drugs or devices. We pulled out the pipes and started to close the chest. I wired the two halves of the breastbone together with the usual stainless-steel sutures, and was just beginning to close the skin when Amo said that her blood pressure was dropping. I stopped closing the wound and stared at the monitor: the blood pressure had indeed dropped from over 100 to 60. Then 50, 40, 30 … and, to everyone's horror, ventricular fibrillation again!

We tried to shock her out of it and failed. There was no choice but to reopen the chest and put her back on the heart-lung machine. I cut and pulled out the stainless-steel wires, and followed the same sequence of frantic steps: internal heart massage, two pipes, on bypass, full flow. Once again, the heart started beating happily within a couple of minutes of the pump being switched on.

Now what do we do?

We considered the possibilities: Emma's heart may just be irritable and with a tendency to go into fibrillation, so we gave drugs to suppress that. The heart may not have had enough 'rest', so we decided to rest it for an hour this time. Is there something wrong with the bypass graft? I dismissed that as a possible cause for two reasons: the first was that I arrogantly 'knew' my graft, had faith in it, and I had just, barely an hour ago, felt it pulsate; the second was that the graft supplied a blocked artery and we knew she had survived fine with no flow at all in that artery, so the graft could only have made things better and certainly not worse. What about her other coronary arteries? Had we missed something? Was there another narrowing or blockage that should also have been dealt with? We carefully reviewed her angiogram and found nothing: apart from the single blocked artery, all her coronary arteries were small but whistle-clean. What on earth was going on?

We resolved that after a full hour's rest we would try again, with drugs given to suppress ventricular fibrillation, and other drugs ready to stimulate the heart should the need arise. Another coffee, badly needed this time as it was one o'clock in the morning.

Once again the heart-lung machine was switched off and the heart took over the circulation without difficulty, but, feeling somewhat paranoid by then, I left the chest open and the pipes in, connected and ready to be used again just in case. We waited another half an hour, staring at the heart, which was beating happily and at the monitors for any sign of a drop in blood pressure or any other deterioration. There was none.

'Come on,' said Amo, 'that's enough hanging around. She's fine. Just close her up and let's all go home.'

I removed the pipes, with a little trepidation, sewed up the holes and began to put the stainless-steel wires in the breastbone. This time I didn't even get the chance to finish the wiring: Emma's blood pressure crashed, followed by ventricular fibrillation. We again took out the wires, started internal cardiac massage and put the pipes back in. We restarted the heart-lung machine.

Something was undeniably happening to this heart. It was behaving as though it was being intermittently deprived of oxygen one minute and fully happy the next. We knew that the coronary arteries were all OK except for one, and that one had what appeared to be a healthy, functioning bypass graft. We simply could not figure it out, but we have a solution to help hearts deprived of oxygen — whatever the reason for that deprivation — and that is a machine called an intra-aortic balloon pump. With this, a sausage-shaped balloon is inserted into the aorta just downstream from the coronaries, and rhythmic inflation and deflation of the balloon forces blood down the coronary arteries.

We had nothing to lose, so we brought in the machine and I inserted the balloon. We switched on the machine and again waited an hour for the heart to rest from the most recent of its tribulations. Again we were able to disconnect the heart-lung machine with ease, but this time I stood there watching the heart and the monitors for over an hour, not daring to take the pipes out and close the chest, lest it should happen again. Finally, I had the courage to go ahead and took out the pipes. This time, I did not even get as far as reaching for the stainless-steel wires: it happened again. Once more I had to put the pipes back in. By then I was running out of space in the aorta and right atrium for the pipes, but I just managed to find places to squeeze them into and Emma went back on the heart-lung machine. Again, the heart started behaving normally.

It was morning by now and the sun was rising. We had been in the operating theatre for over eight hours and still had not figured out what the problem was, let alone found a solution to it. Amo brought in an echo machine, which is able to look at the heart function through an ultrasound probe placed within the gullet. He placed the probe to show the heart beating perfectly, every part of the muscle contracting well, with not one part of it looking even slightly sluggish, which pretty much eliminated any possibility of this being caused by a single coronary artery.

Having had the required period of rest, we tried to come off the machine again, with the echo images running in real-time to see which bit of the heart was the cause of the problem, if such bit there was. We figured that if this was due to a coronary artery, whether it was grafted or not, the echo would tell us which one by identifying the bit of heart that was not working. Again the heart took over the circulation without difficulty and the machine was switched off.

We watched the echo images with intense concentration for about 10 minutes — and it happened again. This time we saw it happen on the echo first, before any drop in blood pressure and before ventricular fibrillation took place: the entire heart started to look sluggish, with every part of it simultaneously beating less and less well, until the blood pressure crashed and the heart went into ventricular fibrillation. That meant that whatever was causing this was something affecting all of the coronary arteries at once, and affecting them equally.

Could Emma's coronary arteries all be going into spasm? Coronary artery spasm is rare, but it does exist and is well recognised. Surgeons sometimes blame it for the poor results of a coronary bypass operation, and are often treated with scepticism. I admit that when I hear a fellow surgeon blame spasm for some crisis that afflicted a coronary patient, my immediate reaction is (rather uncharitably) to assume that the surgeon must have simply botched the operation. Could spasm explain what was happening here? It would explain why Emma had such awful symptoms, despite relatively mild coronary disease. It would fit with the original irritability when she was first put to sleep. It would fit with the fact that things got worse when the heart was touched and interfered with. It would explain why the heart seemed to be globally affected, instead of showing a little bit of local dysfunction related to just one artery. In short, if she was truly liable to coronary spasm, everything unusual about her presentation and subsequent course could be explained.

We fortunately have powerful drugs to counter spasm,

and we immediately set up an intravenous infusion of these drugs. Once the drugs were on board, we tried again. This time, Emma came off the heart-lung machine and absolutely nothing untoward happened. We closed the chest, went back to the ICU and she was fine. By then it was ten o'clock in the morning and I was already late for a training session in which a number of fellow consultants and I were supposed to learn the skills of being a good 'mentor' for newly appointed colleagues. I turned up and sat dutifully through what may have been an extremely useful educational exercise, nodding and smiling at what I hoped were appropriate moments, but in spirit I really was not there. Half of my mind was mulling over the events of the night, while the other half was fast asleep.

Later that year I attended the annual meeting of the McKlusky's Club. This is a gathering of heart surgeons of my generation from around the country. We meet once a year in January to present and to discuss the previous year's disastrous cases in the hope of learning from each other's mistakes. I presented Emma's case and pointed to spasm as the possible cause of all that trouble. I could see that one or two surgeons in the audience were looking at me incredulously, with the implicit subtext of 'Spasm, my arse. He must have botched the bypass.'

Serves me right.

Twelve years later I was horrified when Emma was referred to me again. Her angina had returned with a vengeance and the tests showed that her coronary disease had progressed, so that she now had narrowings in all of her coronary arteries. Needless to say, I was extremely reluctant to operate on her again, knowing the grief that we had the first time round. We tried every possible concoction in the drug cupboard to control her angina, but failed. I did my utmost to dissuade her from having an operation, to frighten her with the thought that if any surgeon so much as touched her heart she was likely to leave the hospital in a box, and with these scare tactics I managed to keep her away from my operating table for an entire year, but in the end she absolutely demanded an operation, regardless of the risk, as her angina had made her life intolerable.

I tried to persuade Jon Mackay to be the anaesthetist, but he declined. We went ahead with John Kneeshaw as the anaesthetist. Under cover of industrial doses of anti-spasm medication, we did two more bypass grafts. Apart from two episodes of ventricular fibrillation, which were relatively easy to treat, Emma sailed through without a hitch. She has had further problems with blood vessels elsewhere, but her heart has been fine ever since, and I fully intend to retire from active practice before she turns up with a heart problem a third time. Last Christmas, I received a card from her, signed: *Emma Chapman, your nightmare.*

There is little doubt that we now know a lot about the heart and the blood vessels, but we certainly do not know everything. Emma's story, and a fortunately small number of others, serve to remind us that there are still aspects of medicine that we do not fully understand. Most of the time, when things go wrong in a heart operation, the surgeon knows, deep inside, what went wrong. There may be a fair amount of denial, rationalisation and justification, but the cause of the problem can be readily pinpointed.

In heart surgery, how the patient does afterwards is very strongly related to how well an operation was performed in the overwhelming majority of patients, but not in all. Once in a blue moon, a perfectly performed operation goes awry in the most unimaginable manner and the rug is viciously pulled from under the surgeon's feet. When that happens, we feel humbled and perhaps a little paranoid. There is always an element of uncertainty in administering treatment to patients. We can and do make massive efforts to reduce the risk of such treatment to vanishingly small percentages, but the risk will never be zero, in part due to events arising from this uncertainty. After such events, we would say to each other that we can never trust cardiac surgery, because just as we begin to feel confident and relaxed about performing it, and after a long period with no significant problems or complications, the specialty rears an aggressive head and bites us on the bum.

CHAPTER 7

World-class surgery on a shoestring

In 1979 I was attached as a fourth-year medical student to a hospital in New England in the United States. I was on the general medical service, and at the end of the working day it was the duty of our team of physicians to go around and see any 'consults', as American doctors called them. That means that patients with medical problems on surgical or other non-medical wards were referred for a specialist medical opinion if they were felt to need the opinion of an expert physician — something that the surgeons looking after them could not provide. Every day a handful of such 'consults' needed to be dealt with before the day's work was done. One day, as part of this 'consult' procedure, we all trouped to an orthopaedic ward to see an old woman who had had a hip replacement. Her problem was that she was a bit short of breath and her chest X-ray looked a tad 'fluffy'. We went to see her, examined her and looked at her X-ray.

'Chest infection,' said one of our physicians. 'Let's prescribe an antibiotic.'

'Heart failure,' said the other. 'Let's prescribe a diuretic.'

'Could be either,' I thought, but being no more than a lowly visiting medical student observer, I kept my mouth shut.

Back in the NHS, a pragmatic doctor dealing with this relatively minor problem would plump for one or the other, start treating accordingly, and if there was no improvement in a day or so, have a change of heart and switch to the other diagnosis or treatment. If the patient was worryingly ill (which this patient certainly was not), an NHS doctor could cover the bases by treating both. This is an easy, pragmatic, safe and cheap approach. Not so in the grand US of A.

After a seemingly interminable argument about which of the two diagnoses was more likely to be the correct one, I was agog to see that the decision they made was to insert a Swan-Ganz catheter to find out for sure if she was in heart failure. A Swan-Ganz catheter is a thin tube inserted into a large vein in the neck and advanced through the heart to the pulmonary artery, which is the artery leading from the heart to the lungs. Once the catheter is in place, a balloon at its tip is inflated and a pressure monitor located downstream from the balloon measures the filling pressure of the (further downstream) left ventricle: if the pressure is high, it is heart failure. If not, it must be a chest infection by default. All of this is rational and pretty scientific, you may think, but inserting a Swan-Ganz catheter is still quite an invasive procedure, requires admission to the ICU (where the monitoring kits are) and is much, much more expensive than antibiotics and diuretics combined. Still, that was the decision of the experts and — somewhat incredulously — I followed the medical team as they escorted the patient to the ICU for the procedure to be done.

One junior member of the medical team was visibly delighted to be charged with inserting the Swan-Ganz catheter: this was clearly going to be a feather in his cap. He scrubbed, put on gown and gloves, set out his trolley and asked for the patient to be laid down flat. He then prepped the skin, placed surgical drapes around the patient's neck and upper chest, and inserted the needle just below the collarbone, looking for the right subclavian vein. Subclavian means 'under the collarbone' or clavicle, and the right subclavian vein is the large vein that drains blood from the right arm back to the heart, and one that is sometimes used for inserting these catheters.

In and out the needle went, with not a drop of blood coming back through it to indicate that he had struck the vein. He tried again, at different angles and at different points below the collarbone: still no blood. He had one more go and, this time, blood shot up the needle on to the drapes. There was a problem, though: the blood was not blue and sluggish, but bright red and pulsating vigorously. This was not the subclavian vein, but the subclavian artery. This is bad news: like the vein, the artery is tucked behind the collarbone and quite deep. Being an artery, the blood within it is at high pressure. Being behind the collarbone, it is difficult to stop it bleeding, because direct manual pressure on it is impossible. Nevertheless, the doctor tried. He quickly withdrew the needle and applied a big wad of swabs above and below the collarbone, pressed quite hard and waited. The patient complained bitterly about the pressure and was given painkillers. After half an hour or so, the doctor removed the swabs and everyone was relieved to see that there was no further bleeding from the artery. He then went on to look for the vein again and, after a few more punctures, eventually found it. He inserted the Swan-Ganz catheter successfully. The patient, meanwhile, was not at all happy: she was complaining bitterly of a sore right arm and was given even more painkillers.

At the end of the procedure, which had taken nearly two hours, the drapes were removed and the assembled doctors and nurses were greeted by the most unwelcome sight of her right arm: it was white and, when touched, very cold. No wonder she was complaining of pain: her arm had lost its blood supply. Clearly, the interference with the subclavian artery and the pressure applied to it had caused a clot to form. The clot had completely blocked the subclavian artery and she was now being threatened with the loss of her right arm altogether. There was nothing for it but to request an emergency vascular surgery 'consult'.

The vascular surgeons came, readily agreed with the diagnosis and recommended doing an emergency 'embolectomy' — or in plain language, clot removal — in the hope of restoring blood flow to the arm, while it was still viable. She was immediately taken to the operating room and put to sleep. The vascular surgeons then made an incision in the hollow of the arm at the elbow, found the brachial artery (the arm artery that continues from the subclavian artery), put slings around it and opened it. They then threaded another catheter with a deflated balloon at its tip all the way up to reach the subclavian artery. Once they felt they had passed the site of the puncture and the clot, they inflated the balloon and slowly withdrew it, bringing out a sizeable sausage of clot. They closed the hole in the brachial artery and stitched up the wound. When the drapes were removed, the arm, hand and fingers were a healthy pink colour and alive again. After celebrating their success and cleverness, they sent her back to the ICU.

The following morning the ICU doctors switched off her sedation and she began to wake up. To everybody's horror, it became quickly apparent that she was simply unable to move any part of the left side of her body. She had suffered a massive stroke. What had happened was this: the attempt to withdraw the clot was only partly successful: most of the clot came out as planned, but a fragment of it was pushed back by the manipulations. 'Back' from the subclavian artery is the origin of the right carotid artery, which supplies the right side of the brain, and that is where the clot went. The right brain controls the left body, hence the left-sided paralysis. The hapless woman was still in ICU when I left to return to England at the end of my two-month stint, and I do not know if she survived her medical misadventures, and if so, what kind of a life she had afterwards.

And the Swan-Ganz? It had shown no heart failure, so all she had needed was antibiotics. Having added diuretics to that would have cost an extra dime or two.

The remarkable feature in this story is that such a complication would never have taken place in a National Health Service hospital in the UK, where treatment decisions are often guided by pragmatism and with an eye on the likely costs and invasiveness of procedures. Let me now tell you a bit about the NHS, because it is truly an amazing organisation, despite the constant scare stories and negative media coverage that it receives on a weekly and sometimes even daily basis. In fact, if you are a reader of the *Daily Mail* and one or two other newspapers, you could be forgiven for thinking that the NHS is a dysfunctional and decrepit system populated by careless and incompetent staff hell-bent on doing harm to patients. Nothing could be further from the truth. Of course, mistakes are sometimes made, and of course awful things happen every now and then. As in most large organisations, there are a few 'bad apples'. There are people who are not up to the job, lazy workers and others who simply do not care enough about the quality of service that they deliver, but, by and large, these are the minority, and probably a smaller minority than that which exists in many other large organisations, be they public or private.

The NHS delivers a pretty good standard of health care from cradle to grave without requiring any significant form of payment from the user at the point of care, and it does so very well. Everything from the contraceptive pill to a coronary bypass is available to you and you do not have to prove an ability to pay before being treated, which means that, if you are suddenly taken ill, you can go to the Accident and Emergency Department without your wallet or credit card. Most of the workers in the NHS are salaried and are not paid by procedure or by item of service. Because of that, unnecessary treatment rarely takes place in the NHS. Compare that with the United States — where everything done to a patient can potentially bring hard cash to the doctor that does it and to the hospital that provides the setting — and you can see the direct and unrelenting monetary pressure to do something rather than to do nothing. Add to that pressure the ever-present fear of litigation in case something is missed or not acted upon immediately and you can understand the irresistible drive to over-investigate and over-treat, which can so easily lead to catastrophes like the one I witnessed in Connecticut. Most investigations carry some risk, even if that risk is merely a small amount of unnecessary radiation, but the risk of over-investigating does not stop there.

There is an apt aphorism in medical investigations: 'Nobody is normal: they just have not had enough tests.' If you subject yourself to every conceivable blood test, ultrasound, X-ray, magnetic scan, nuclear scan, endoscopy, electrical recording, swab, smear and scrape, then something is invariably bound to be unearthed and found to be abnormal, even if you are perfectly happy and healthy.

What is even worse than the above is the fact that, whenever a test is carried out, there is always a possibility that it can actually identify as wrong something which may not be wrong — and it may miss something which is not right. When a test shows that 'something is wrong' which isn't, that is called a 'false positive'. Similarly, a test missing something wrong which is actually there is called a 'false negative'. False negatives, apart from the inherent risk of the test itself, are no worse than not having the test done at all, but false positives — even when the risk of the test itself is small or non-existent (such as ultrasound or magnetic resonance imaging) — inevitably lead to additional tests, interventions and operations to sort out the thing thought to be wrong, but which was not there in the first place, and the size of the risk begins to pile up alarmingly. That is why screening can be bad for you.

Liam Hughes, the Norwich cardiologist who referred Emma Chapman, was once discussing a patient who had gone through a very large number of increasingly invasive procedures, because of some probably benign finding picked up at a heart-screening test. After listening to the case being presented, he declared that in his opinion the patient was a VOMIT. We all looked at him, perplexed. He then explained that he had coined the term to describe patients who were 'Victims of Medical Investigation Technology'. The NHS, in comparison with the health-care systems of other prosperous countries, is a conservative system with relatively little in the way of VOMIT. Much of the rest of the world sees NHS care delivery as rather slow, not sufficiently proactive and with a tendency to under-investigate and under-treat. This may well be true, but at least events like those that befell that poor old lady in my US student elective do not often happen here.

The other remarkable thing about the NHS is its staff: there is a degree of pride in the professionalism of all NHS staff, be they nurses, doctors or paramedical professionals. Most NHS workers stay to finish the job without clock-watching. They generally work much harder and longer than they are paid to do. They feel a duty towards their patients that few service providers have towards their customers, and they gain enormous satisfaction from doing a good job and doing it well. The sense of duty that we have towards our patients sometimes borders on the ridiculous: most NHS heart surgeons that I know are on call for their own patients 24 hours a day, seven days a week. They do not hand over their patients' care, except when they are physically out of the country, and even then, they will call often to check on their patients' progress and have input into their management from all corners of the globe. Of course, the ubiquitous roaming mobile phone has made that task even simpler, and we are now almost never 'off duty'.

A few years ago I was asked to speak at a conference in Sydney. I was there for only three days, which passed in a blur of jet lag. One of my lifetime ambitions was to sail under the Sydney Harbour Bridge and, as I was talking about that with other delegates, one Australian anaesthetist said he had a big yacht moored nearby in the harbour area and asked,

'Would you like to do that tomorrow?' Thus it was that on a gloriously sunny day, with a gentle wind propelling us past the Opera House, he handed me the helm of his beautiful 50-foot sailing yacht, and we slowly approached the bridge. Just as we were passing under it, my mobile phone rang: 'Hi, Mr Nashef. It's Dr X from Papworth. Mrs Brown's leg wound looks much better now. Can she go home tomorrow?'

Despite its monolithic and unwieldy nature, the NHS somehow brings out in its staff a level of dedication to duty that is second to none, and if the NHS is truly the jewel in the crown of the British way of doing things, then my hospital is the jewel in the crown of the NHS. So now, let me tell you a bit about Papworth. Its physical location and appearance could not be less impressive: it is housed in a randomly scattered group of low-rise buildings, a ramshackle arrangement of old Victorian, red-brick structures interspersed with all shapes, sizes and colours of prefabricated Portakabins and the occasional, slightly more modern, functional, purpose-built block. A narrow road with frequent and vicious speed bumps snakes through it, and at one edge of its grounds is the famous duck pond, home to a variety of ducks and coots, the occasional black swan, some large carp and a regular visiting heron, who presumably feasts on them. The hospital is in a little Cambridgeshire village, which started life as a small residential community containing the headquarters of a Trust that looks after many disabled people and offers them employment and support. The hospital itself and the village within which it resides are so underwhelming that on my first visit (to be interviewed for the job of consultant surgeon) I arrived, looked around and was absolutely convinced that the taxi driver must have delivered me to the wrong address. And yet, from this unpropitious setting, some of the most amazing, world-leading developments in heart surgery have emerged over the past three decades.

This hospital, housed in buildings which look barely fit for a Third World nursing home, carries out more than 2,000 major open heart operations per year, the largest number by far of any heart surgery unit in the UK, with risk-adjusted results that are a 'positive outlier', which means the survival rate is statistically better than the nation's average, despite operating on the sickest and oldest patients in the country. On top of that, there is an operation called pulmonary thrombo-endarterectomy (or PTE) — developed mostly in San Diego in California — to treat patients whose lungs become progressively clogged up by clots flying in from elsewhere in the body. These patients are desperately breathless and will die soon unless their lung arteries are cleared out. It is a laborious, technically demanding and difficult operation, which usually takes all day to complete. Papworth performs more PTEs than any other hospital worldwide, again with excellent survival rates. We also do more heart and lung transplants than anywhere else in the UK with similarly excellent survival rates.

We have no formal academic or professorial departments and the overwhelming majority of our doctors are full-time jobbing clinicians. Despite that, we produce an astonishing amount of research relating to medicine and surgery of the heart and lungs, and have the best training programme in the country, if not the world, for aspiring heart surgeons. We have a string of firsts, including the first successful heart transplant in the UK, the first successful heart-lung transplant in Europe, the first minimally invasive coronary bypass in the UK and many others.

In short, this is a very good hospital, worthy of nurture and support. And yet it does not receive the support and help it needs to develop and expand. After decades of trying to rebuild the hospital so that we have decent facilities, we are now finally moving to a new hospital in Cambridge. You would have thought that, having proven our worth to the nation's health service, and having produced, against all odds, such fantastic outcomes from so inauspicious a setting and for so long a time, the Department of Health would have provided us with a generous capital investment to build a state-of-the-art facility from which we could continue to serve and improve. Sadly not. We are largely funding the building ourselves, through a Private Finance Initiative. The new design is indeed a very pretty structure, but, unfortunately, not big enough for the work that we do, want to do, and need to do to secure the income to pay for it. The new building is all we could afford and many compromises have had to be made. Papworth may end up financially broke, and if the austerity drive that is curtailing NHS spending continues at this pace, this could happen sooner than we think.

The NHS, as I have said, is a wonderful organisation,

but its main problem was always — and still is — a lack of money. For one of the top economies in the world, the UK simply does not spend enough on the health of its citizens. Successive governments of different political persuasions have consistently tried to brush that fact under the carpet, while pretending to improve the health service of the nation by achieving greater 'efficiency'.

At Papworth we use our operating theatres seven days a week, including Saturdays and Sundays, for routine elective operations. On weekdays, the theatres routinely run till eight o'clock in the evening, and often beyond, with elective surgery. If an emergency operation is needed, we often have to cancel an elective operation to deal with the emergency. Most hospitals manage two major heart operations a day per operating theatre, but we often squeeze in three. The occupancy rates of our intensive care unit hovers around the 100 per cent mark, which means that we frequently have to cancel operations when something unexpected happens and the beds in the intensive care unit are fully occupied. We discharge patients from the intensive care unit as early as humanly possible to make room for the next batch. Sometimes we discharge them a little too early, so that they 'bounce back': we have the highest rate of readmission to ICU in the country. Our surgical wards are so full that we often admit patients for major heart surgery on the actual morning of their operation to save on bed occupancy. While these patients are in theatre or ICU their beds get used for other patients in a process called bed-hopping. There is no rattle room anywhere. Members of the clinical staff work very long hours and most of them work far harder and much longer than they are paid to do.

Dear Health Minister: how much more 'efficient' would you like us to be?

CHAPTER 8

The story of A and H

People usually assume that doing heart surgery is difficult. It is not. The technical side of the job is no more than a set of skills and, like all skill sets, it can be learnt. With enough practice and guidance, almost any doctor who is not totally cack-handed can become a technically competent heart surgeon. The part of the craft of heart surgery that is most demanding and most difficult to acquire is not the cutting and stitching, but the decision-making. That — even more than learning the different steps of various operations — is the reason it takes a decade of training to become a heart surgeon. Decisions have to be made all the time, from the planning of an operation in the cold light of day during a clinic visit all the way to decisions which have to be made 'on the hoof' when the unexpected happens during an operation. One of the most critical decisions that surgeons make on a daily basis, well before a patient is anywhere near an operating theatre, is whether or not to operate at all. As much harm can be done by an ill-advised operation as by not operating at all when surgery is truly in the best interests of the patient. Sometimes we only know what should have been the right decision after the event, when the wrong decision may have been already made and it is too late to do anything about it. John Wallwork, who was, until he retired a few years ago, a very good cardiac surgeon at my hospital, used to say at the zenith of his career that he was technically no better than the senior trainees who worked with him. He often remarked that 'The only difference between senior trainees and me is that I make better decisions than they do.' Decisions are the stuff that the surgical profession is made of, and the wrong decision can have an impact as catastrophic as a botched operation.

Wherever possible in this book I have tried to give, freely and openly, the real names of real people. Where I have done so I have, of course in advance, sought the permission of the patients concerned and, whenever that permission was not forthcoming or not obtainable for any reason, I have changed the names to protect identities and thus comply with my duty to respect patient confidentiality. Sometimes, due to issues related to confidentiality and a few other reasons, it may not be a good idea to give the names upfront, so the next two cardiac patients I shall tell you about will be known as A and H.

A and H were a married couple. A was a United Nations official and he worked in education. H was a writer passionate about classical and modern literature. They were both smokers and did not look after their health particularly well. They led relatively sedentary and intellectual lives and, in addition, A was slightly overweight and had very high blood pressure, so it was perhaps not a great surprise when, while still in his early fifties, he woke up one morning with crushing chest pain. He tried hard to dismiss the pain as a bad bout of indigestion, but no indigestion remedies worked and the pain just got worse. He and his wife both realised that this could actually be serious and she called for an ambulance. He was immediately whisked to the local university hospital, where a massive heart attack was diagnosed.

The heart attack had left him with a severely damaged heart with poor function of the left ventricle, which is the heart's vital pumping chamber. For over a week he was teetering on the brink between life and death, with dangerous heart rhythm disturbances every now and then and quite severe heart failure. Eventually, he began to rally. His heart failure gradually started to recover and his rhythm disturbances looked like they would abate with the use of powerful heart rhythm stabilising drugs. When he was finally deemed fit enough to be discharged, he went home, a diminished man, weakened and tired, with a long journey to anything resembling a full recovery still ahead of him.

Sometime later, he had a coronary angiogram to look at what shape his coronary arteries were in and this showed that the situation was pretty bleak. He had severe coronary disease affecting all of his coronary arteries, and not just the one that had been blocked when he had his heart attack. Normally, this alone would be an indication for a CABG, but his left ventricle was found to be awfully weak and, to make matters worse, he had a left ventricular aneurysm, which means that the area of the heart that was damaged by the heart attack had thinned out and ballooned, adding more stress to the weakened heart's ability to pump blood effectively.

Such patients, if they present themselves nowadays, would be offered a CABG, plus removal of the aneurysm, at a slightly higher risk than would be expected in a standard CABG alone, but at a risk which is in no way prohibitive. Unfortunately, however, this was in the 1970s, when cardiac surgery was still a relatively young specialty, and the risk of operating on A's heart was unanimously deemed to be too high to contemplate. A consulted many cardiologists and cardiac surgeons, including prestigious ones in Harley Street, and was advised by all of them to continue with medical treatment and to avoid surgery, and this served him well for about a decade.

Nowadays most of us heart surgeons are, if anything, even keener to offer CABG to patients with a poor left ventricle, despite the fact that the risk of such surgery is higher. The reason for this is simple: patients with coronary disease usually die of heart attacks. If the patient has a poor left ventricle, he or she can ill afford to lose any more heart muscle through yet another heart attack. For such patients, the next heart attack is more likely to kill them than it would a patient with a strong, healthy left ventricle. In simple terms, having a poor left ventricle means that you have no 'reserve' left: your next heart attack will therefore probably be fatal. Of course, CABG in such patients carries a higher risk, but it also helps to prevent further heart attacks, massively reducing the future risk of the disease. That is one reason why the often quoted 'heart surgery paradox' is true: *the more an operation is likely to kill you, the better it is for you.* A patient with a good ventricle has a relatively low risk for CABG, and also a relatively low risk from being left alone without surgery. A patient with a poor left ventricle has, it is true, a higher risk from surgery, but a far, far higher risk from being left without an operation.

This simple fact was either not known or not yet true in the specialty as it was in the 1970s and so A, despite having a weakened left ventricle and severe coronary disease, was left without surgery and, of course, the inevitable happened: he had his next heart attack about 10 years later, and this one was indeed fatal. He died at the relatively young age of 62.

The story of his wife, H, is the diametric opposite. She developed heart disease at a more advanced age and in the 1990s. She had tests which showed a narrowed aortic valve and some moderately tight narrowings in two of her coronary arteries. Unfortunately for her, this relatively modest heart condition was not easily tolerated either by her or by her heart: she kept going into heart failure. Her doctors offered her medical treatment using a variety of different drugs, but whatever treatment was dished out to her was either ineffective or poorly tolerated or both. She had been admitted to hospital with heart failure for the third time in the space of a few months and, on this occasion, the cardiologists decided that they had no further medical treatment to offer. They therefore referred her to the cardiac surgeons. The surgeon who saw her was admittedly a tad concerned that her symptoms were a little excessive and somewhat out of keeping with the extent of her heart disease, but he too could see that there was no option but to offer her an operation if she was going to be able to leave hospital at all, and so an urgent operation was planned to replace her aortic valve and perform a CABG for her obstructed coronary arteries.

The operation did not go well. Her heart struggled to take over the circulation from the heart-lung machine at the end of the procedure and needed support from powerful, heart-stimulating drugs before it could do so. She was moved to the intensive care unit in a critical but stable situation. Over the subsequent few hours her heart function deteriorated further, and higher doses of drugs were needed, but the inexorable decline continued. Two of her children, a son and a daughter, were themselves doctors. Her son was by her bedside when her heart stopped. He did his best to help the ICU nurse who was attempting to resuscitate her, but to no avail. H sadly died having never woken up from her heart operation.

This story is poignant for two reasons. The first is that a married couple have both died through cardiac surgery in completely different ways. One of them died as a direct result of surgeons being too scared and too conservative to offer him an operation, and the other died through complications arising from a relatively simple operation that was only too readily offered to her. Heart surgery can of course kill, and it can do so both through commission and through omission. With hindsight, it is easy to say that the wrong decisions were made, but at the time all concerned did their best to make the best decisions in the interests of both of them. Nevertheless, it is difficult to escape the feeling that here was one couple who sought help from my own specialty of cardiac surgery and, whichever way you look at it, the final result was that cardiac surgery failed them both.

The second reason is perhaps rather more poignant: A and H were my parents.

CHAPTER 9

When the pump is broken

Yesterday was the 15th of December and I received a Christmas card with the usual festive wishes. It was signed: *David Round and family — heart transplant, 24 years ago*. I have received a similar card from David every year for the previous 23 years and, as every year, it was an emotional moment which brought a smile.

Heart transplants are at the very glamorous end of cardiac surgery, and people outside the profession are usually very impressed by them. There is no doubt that a successful heart transplant is a triumph of all sorts of endeavour, especially in the science of the immune system and the logistics needed to make the procedure go smoothly, but the technical aspects of actually performing the operation itself are not too demanding. In fact, doing a heart transplant is remarkably easy, especially if the patient has not had any previous heart operation. Easy as it is, it remains visually and conceptually very dramatic.

An important feature of the conduct of a transplant that may not immediately spring to mind is that this is not one operation, but two, usually conducted at separate hospitals far away from each other and sometimes hundreds of miles apart. Before a 'new' heart can be put into a recipient, it first must be taken out of a donor who is legally 'dead'. That means someone who has had a horrific accident or brain haemorrhage or some other awful condition that has damaged the brain irrevocably, so that as a result any form of conscious existence in the future has become impossible. In the UK this is determined by a set of 'brain-death tests', which establish that the brain is no more. These tests must be conducted with no sedative drugs in the donor and should all be repeated a few hours later, before the donor is declared brain-dead and his or her organs may be legally offered to be transplanted into someone else.

Of course, the heart is not the only organ that will be used in this way: the lungs, liver, kidneys, pancreas and gut may also be transplanted, and they may be accepted for use in recipients in several different hospitals; so that the donor operation will be conducted by a number of teams from various transplant services, who will almost certainly have to travel to wherever the donor is located from different parts of the country. The arrangements for the removal of organs from a donor pose quite a logistical challenge and are coordinated by a central body in terms of timing, team arrivals and departures, and the technical requirements for the procedure, which differ between organs and may also vary between two teams interested in one organ.

All of this has important implications, especially for the correct time to start the actual transplant operation in the recipient back at the base hospital, such as Papworth. A typical scenario is that some poor soul is admitted to hospital somewhere, perhaps during the weekend or at night, after a tragedy, such as a horrific car crash or a brain haemorrhage that happened out of the blue. The usual immediate care and resuscitation are given and the patient is admitted to the local intensive care unit. The following morning, the patient is reviewed carefully and the possibility that the brain is damaged beyond repair is raised. A set of brain-stem death tests is done and confirms the worst, and another set is done a few hours later. There is now no doubt left: the patient is brain-dead.

When that is ascertained, the family members are told the bad news. An appropriate time interval allows the family to begin to come to terms with the process of grief and bereavement. After that, a formal approach is made which raises the possibility of organ donation. If approval is forthcoming, the transplant authority is informed. The organ or organs are then offered for transplantation to the various transplant centres around the UK in a strict order based on whose turn it is. If the first centre has a patient on the waiting list who matches the donor for size, blood group and other essential criteria, the centre will accept the offered organ. If not, the offer goes to the next centre in the queue, and so forth until all the organs have been found a home. As that process is going ahead, patients on the waiting list for a transplant will receive a phone call inviting them to come to their transplant hospital: 'Mr X, we may have a new heart for you. How quickly could you come to Papworth?'

By now it is probably mid-afternoon on the day of the brain-death diagnosis. The donor teams in several transplant centres begin to prepare for the trip to the local hospital, which is housing the donor, and an operating theatre is organised in that hospital for the donor operation to be performed. That time is likely to be about eight o'clock in the evening at the earliest. The teams arrive, check all the tests on the organ that matter to them and if they are happy with what they see, coordinate the steps of the donor operation between them.

Sometime between nine and ten in the evening, knife is put to skin in the donor. Over many years of doing heart transplants, I have learnt one consistent feature of the timing of transplants: no matter where the donor is and regardless of how slick or efficient the donor teams are, from that point onwards it will be roughly about four hours before the heart arrives at Papworth. This time period will include the removal of the heart, the packaging of it in protective solutions on ice, and the transport back to Papworth (sometimes by car, others by private jet, depending on whether the donor is in Cambridge or in Dublin, for example). Based on the four-hour rule, I usually request that the recipient patient at Papworth is ready and asleep an hour earlier. What this means is that the majority of heart transplant operations tend to start sometime between midnight and one o'clock in the morning. The transplant team is usually a tired group of people starting work at a time when ideally they should be going to bed. When seen in that light, it is perhaps just as well that a heart transplant is technically not too demanding an operation!

Most of the time everything works as planned and the new heart arrives just as the old one is being cut out, but a process involving multiple surgical teams from different hospitals around the country — and various modes of transportation to and from these hospitals — contains within it every possibility of things going wrong. Recently, Evgeny Pashluvkov, who is one of our senior transplant trainees, was out on a 'donor run' to retrieve a heart and things did not quite go according to plan. Here is an account of his journey back to Papworth in his own words:

> We retrieved the heart in a hospital in Middlesex and were carrying it back in the evening to Papworth in our transplant retrieval ambulance, which is essentially a van that is equipped with blue flashing lights and sirens. The motorway was very busy, as it was rush hour. We were already a bit behind schedule due to the heavy traffic and had to use the hard shoulder on several occasions to make up for lost time (in retrospect, that could have cost us a tyre, since there is always a lot of garbage lying on this part of the road). It was my fourth donor

run in a row and my last on-call in a stretch of shifts, so I was quite tired and looking forward to coming home and having some rest at last.

As we were passing Cambridge and joining the A428 dual carriageway towards Papworth I heard a very strange noise coming from somewhere underneath our van. I asked our driver Steve whether he could hear it too and he said, 'Yeah …' rather tentatively. Then I asked him again whether it might be the sound of a flat tyre and he answered with the same degree of equivocation: 'Maybe …' I then asked a third question as to whether we could make it back to Papworth with a flat tyre. This time Steve sounded even less certain: 'I don't know, to be honest.'

It was not too long before the van started to handle somewhat erratically and lose speed. There was no option but to pull up on the hard shoulder and stop. We jumped out of the van to find the cause of the trouble: one of the rear tyres was completely flat. The acrid smoke and smell arising from the wheel clearly indicated the van was not driveable in its current state. We had a quick chat regarding the best plan of action for this unprecedented situation we found ourselves in. The clock was ticking and I was not quite sure how long it would take us to change the tyre in the dark on such a big and heavy van or whether to ask the police for help and wait for them to rescue us. All I knew for sure is that there was a patient lying on the operating table, on a heart-lung machine and with his heart about to be discarded into the bin, and that any minute of delay would be costly for that patient. I had never been in that situation before and there was no chapter entitled 'Flat Tyre on the Donor Run' in our transplant management guidelines. But I knew that I had to get this box of ice with the heart in it to Papworth as quickly as I could.

I decided that the quickest way to get the heart to the hospital was to hitchhike. I jumped out on the road and started to wave my arms like a madman. Steve Fakelman, our donor-care physiologist, was doing the same to increase our chances of success in this unusual undertaking. It was night-time and we were not wearing reflective clothing. The passing cars were not even slowing down, probably because most of their occupants could not even see us.

Eventually one car stopped and a gentleman in his forties inquired what he could do for us. I explained that I had a box with a human heart in it that needed to be delivered to Papworth Hospital as quickly as possible. At these words the gentleman started to look slightly anxious, said he could not help us and sped off into the night. The next vehicle to stop did not keep us waiting for too long. This time it was a massive heavy goods lorry.

I took my time to climb the steps leading to the cabin and tried to use my best English to convey the 'heart of the matter' to the lorry driver. He said that he was not allowed to change the route and divert to

Papworth. I promised that I would sort everything out with his employers and he should not worry about it. He eventually agreed, although not without a great deal of hesitation. It took some effort from Steve Fakelman and me to load the ice box up into the cabin. My retrieval team mates stayed with the stricken ambulance van and I climbed into the passenger seat of the lorry.

We started what I thought would be the final leg of our journey back to Papworth. I felt quite relieved and even the relatively slow speed of 55 miles per hour could not spoil the moment. My gently expressed concern regarding the speed we were driving at was met with a phlegmatic response from the lorry driver: 'This is a lorry, mate, not an ambulance.'

The pleasure of making good progress in the lorry did not last for very long, though. The lorry started to slow down and stopped altogether. The countless red dots of brake lights of the cars in front of us left little doubt as to the cause of the standstill: a major traffic hold-up. It was very long. I could not see the beginning of it — just an endlessly winding red chain of brake lights as far as the eye could see. It was a very depressing view. There were just two of us: the driver and myself sitting in the cabin with the heart in an ice box between us. We were gridlocked by the cars in front of us and, now, behind us. I had not the faintest idea what to do next.

I then saw, somewhere far ahead and very faintly, the flashing blue lights of an emergency vehicle (ambulance? police? both? I could not be sure, as it was quite far away from where we were). It looked like the emergency vehicles were on the other side of the road. Only then did I realise that the opposite carriageway was empty. It must have been closed to traffic to facilitate the work of the emergency services. I decided to seek help from whomever those blue lights belonged to – there was nothing else to do. Before disembarking, I asked the lorry driver to switch on his cabin lights so I could have a chance of locating him in the cabin among the many other lorries also caught up in the traffic jam.

I jumped out of the lorry and ran in the direction of the blue lights. While running, I looked back briefly to memorise the appearance and location of 'my' lorry in the queue. My heart sank at the thought that I was leaving the heart somewhere in the cabin of some lorry in a long chain of vehicles in the middle of the night. I was afraid of even considering the possibility of not being able to find my lorry with its precious cargo on the way back. At that stage, it was too late to do anything differently — I was committed.

When I finally reached the blue lights it turned out to be a traffic-accident scene: a motorbike was lying on the road and a police car and two officers were just next to it. I approached one of the police officers and, trying to catch my breath, started to tell the story of the heart and a flat tyre for the third time during this evening. The police officer looked at me with a mixture of disbelief and suspicion. He then enquired of the whereabouts of this heart I was talking about. I waved in the direction of the traffic backlog, explaining that the box with the heart is somewhere there in a lorry distinguished by having a lit cabin. I could see it in his eyes that my story did not sound plausible for him and he added that he could not really leave the scene. However, I had nothing to lose and I put my remaining energy into words explaining that it was a matter of life and death to get me to hospital with this box. Eventually he gave in.

I asked him to wait for me and ran back to my lorry, praying to find it where I had left it. Fortunately, the lorry was still there, and the plan of keeping on the lights in the cabin worked. I thanked the driver, who helped me to unload the box down from the cabin, climbed over the central barrier on the road, and ran with the heavy box all the way down to where the police officer was waiting for me. We put the box on the back seat of his BMW estate and made our way to Papworth at speeds which far exceeded the limited capabilities of the lorry.

As soon as I saw Ann Thomson, our transplant coordinator, waiting for us at the very entrance of the hospital I realised that nothing could now prevent me from delivering this heart to the hospital any more. The police officer noted that it was the first time in his career that he had been 'hijacked while on duty'. From my side I reassured him that it was not, by any means, the standard way for us to get the retrieved organs to hospital.

I delivered the heart to the operating theatre 15 minutes later than the initially estimated time of arrival. The recipient got his new heart and did very well.

I have used different types of transportation on donor runs before and since that night, but never again a combination of an ambulance, heavy goods lorry and a police car.

The lives of heart donor teams abound with stories like the one above. During my own stint on the donor team in my youth I experienced terrifying speeds in police cars, driving on the wrong side of the road in rush hour, turning up at the wrong hospital, one particularly hair-raising motorcycle police escort, and one ridiculously memorable time when, through a misunderstanding at Manchester Airport, my team and I, laden with surgical instruments, were made to fight our way through the burger joints and duty-free shops at the airport to reach the small jet waiting to take us to Dublin to fetch a heart. Other donor teams have had even more dramatic escapades, including a plane crash (in which all passengers, fortunately, survived and with the box containing the heart intact).

While all this drama is taking place on the donor side, the recipient, oblivious to it all, is being prepared to receive the new heart at Papworth. Here is how we do a heart transplant, back at base and away from the trials, tribulations and transportation troubles of the donor team.

First, we cut through the sternum or breastbone with a power saw. Then we follow the usual early steps of most heart operations. To put the patient on the heart-lung machine, we insert a pipe in the aorta and two more into the two big veins that deliver the blood to the right atrium, rather than just one in the atrium itself. We then encircle those two pipes with tapes snugged tight to isolate the veins from the right atrium. The reason for this minor variation is that it allows us to cut the right atrium without letting air enter the pipes and clog the flow through the heart-lung machine with an airlock. Once the pipes are connected we start the machine and clamp the aorta, isolating the heart from the circulation.

This time, of course, there is no point in protecting the heart with a cold potassium solution, for the simple reason that this heart need never beat again to keep this particular patient alive. We then cut out the heart across four structures: the two atria (the way in to the heart) and the two big arteries coming out of the heart (the aorta to the body and the pulmonary artery to the lungs). While we are doing this, the heart is slowly dying of oxygen starvation. One of the saddest sights in heart surgery is seeing the poor, condemned heart as it desperately but determinedly continues to beat, more and more weakly, while it is being cut out. It usually carries on beating ever more feebly for a few minutes, even when it is sitting, discarded, in a metal tray on the instrument trolley. Then it beats no more. Meanwhile, there is a huge space in the middle of the chest between the lungs with nothing but three pipes coming out of it. Provided the logistics have all gone according to plan, just as the old heart is coming out, the new one is being carried in to the operating theatre in a large box containing ice, within which there is a bag containing cold saline, within which there is another bag containing the heart itself.

Plumbing in a new heart needs only four stitch lines, and none of them is especially delicate: left atrium of the new heart to the remaining cuff of the left atrium of the old heart; right atrium in a similar fashion, pulmonary artery of the new heart to pulmonary artery of the recipient, and similarly the aorta. Of course, all of these suture lines need to be blood-tight, like all joins of blood vessels, but that is well within the heart surgeon's abilities. The stitching, however, takes a long time and the right atrial suture line is the most tedious. Once all the joins are made, any air in the new heart is washed out and the clamp across the aorta is removed.

With the removal of the clamp, blood flows down the new heart's coronary arteries and two things will simultaneously happen, the first of which is truly magical: after being removed from the donor and after spending hours in cold storage, the heart will start to beat in its new home. The second is rather less desirable: the recipient's blood cells will instantly recognise this heart as foreign, and the primed immune system will immediately set out to kill it. To prevent that, a powerful cocktail of immune suppression drugs is given as the recipient's blood and donor's heart make their acquaintance for the first time. Then the heart-lung machine is switched off and disconnected, the chest is closed and it's all over.

Over the next few days the patient walks a tightrope: too little suppression of the immune system can lead to rejection and damage to the new heart, and too much of it makes the patient vulnerable to infections elsewhere. After years of transplantation, we have teams of people who are very experienced in helping the patient negotiate that tightrope successfully by judicious use and fine adjustment of the medication against rejection and infection. A time-honoured tradition at Papworth is that when the patient with a new heart can walk round the duck pond, he or she is ready to go home.

Stepping a little back now, let us look at the bigger picture, which provides the context for the amazing procedure that is a heart transplant. Heart surgery was born when the heart-lung machine was invented in the 1950s and has grown at a frantic rate ever since. Over the last half century the range of heart problems that can be fixed with a knife and a stitch has expanded massively, so that we surgeons believe and behave as though there is nothing in the heart that we cannot fix. We can bypass blocked coronary arteries, replace or repair all four of the heart valves, replace or repair any problems with the aorta, close holes in the heart where there should not be holes and create holes where there are none but there should be. Other ingenious surgical solutions have been devised to fix any faulty plumbing that a baby may be born with, including plumbing defects which are almost impossible to comprehend, yet most certainly exist, such as a missing ventricle, missing major arterial and venous connections, and even ventricles and atria connected the wrong way. More recently, surgical solutions have been found even for electrical rhythm disturbances of the heart, such as atrial fibrillation, where the electrical signals in the atria go haywire and give the patient palpitations, a less efficient heart and a higher risk of stroke. Even that can now be fixed with an operation, so that we, the plumbers, have now also become electricians.

Unfortunately, there is still one heart problem which manages successfully to defy the heart surgeon's knife, and that is heart failure. Of course, if the heart is failing because of coronary or valve problems, we can usually fix it. However, if the heart fails because the pump itself is no good, it is a different story. There is simply no operation that can be done on the weak and failing muscle pump itself to make it beat better and pump harder. Pump failure is bad news, and heart failure in general is responsible for many deaths, with 30 to 40 per cent of patients dying within a year of diagnosis, making it a far worse disease to have than cancer.

Why should the pump fail? Reasons are many. The first is coronary disease: when a coronary artery gets completely blocked, a little piece of the heart muscle dies (that is what a 'heart attack' is). Every time this happens, the heart loses more and more of its pumping power, until it can become so weak it just cannot pump enough blood around the body. We can bypass the blocked coronary arteries to prevent any future heart attacks, but that will not bring dead heart muscle back to life.

Other things can also cause heart failure: a tightly narrowed or badly leaky valve makes the heart work very hard to keep the forward flow going, despite being constantly hampered by the narrowing or the leak, so that the heart is working excessively all the time. To the heart, the amount of pumping effort needed is as though the patient is running a marathon, when in fact he or she is sleeping or resting in an armchair. A heart can do this for a short while, such as for the time needed to run a marathon, but not for weeks, months or years on end. Sooner or later, if the valve problem is not fixed, the pump will fail and the failure may be irreversible.

Other occurrences are rarer, but can still happen, such as viral infections or disorders of the immune system that directly damage the heart muscle itself. Whatever the reason, when the pump itself fails, surgical options become very limited indeed and the outlook is dismal. At the moment we have no method of regenerating damaged heart muscle. A large amount of research is currently going on to see if that can be achieved using stem cells, but this, so far, has not been rewarded with any degree of clinically useful success. For the time being, when the pump fails, the only surgical option is to replace it with a new one, and that is where heart transplants come into the picture.

Heart transplants, glamorous and exciting as they may be, represent therapy which is far from perfect, for three major reasons.

The first is the one we have already discussed: the new heart is foreign to the immune system and will be rejected unless the immune system is suppressed with drugs. When the immune system is thus compromised, infections become a problem and so do some rare cancers, especially those of blood and bone marrow. The end result of all of this is that the disease of having a failing heart is replaced by one of having a transplanted one: better than the old condition, but still one with problems which will inevitably become worse over the years.

The second is that there is no endless supply of donor hearts: it is a resource which, by its intrinsic nature, is very severely limited. Heart transplantation is therefore only for the select few, who tend to be those patients with a failing heart who are not too old and have little else wrong with them: this is not being ageist or discriminatory in an evil way, but merely an attempt to eke out the maximum benefit from a limited resource. Even with these restrictions, there are simply not enough hearts to go round. This means that many patients with heart failure who would benefit from a heart transplant may not get one, and may die on the waiting list before a suitable one becomes available. When heart failure is so common and afflicts so many, it is hard to see that a limited treatment such as transplantation can do much to alleviate the problem, when it can only be available to a chosen few. There is an aphorism which is frequently quoted in many forms, but its essence is this: *heart transplantation is the cure for heart failure in the same way that the lottery is the cure for poverty.*

Finally, there is an uncomfortable truth in heart transplantation which is shared with the transplants of many other organs: it is a treatment that requires someone young and healthy to die so that someone old and sick can live. Behind every transplant triumph, there must be a human tragedy.

There is a solution on the horizon. Since the heart is merely a pump which can deliver five litres of blood a minute, it should not be beyond human ingenuity and engineering to make a replacement. Work on this has been proceeding for many years now, and the devices are improving rapidly. Pumps which can keep a patient going for weeks with relative safety until a donor heart is found are now commonplace in transplantation. We call these a 'bridge to transplant' and they have allowed hundreds of patients to survive until a donor heart was found. Pumps designed to work for life (called 'destination therapy') have also been made and used in many patients. They are not perfect and there are problems with what they do to the blood and with their power supply, but they do work. It is only a matter of time before they become better than transplanted hearts and are available on the shelf of every cardiac operating theatre. When that happens, heart transplantation will die out and make way for heart replacement.

CHAPTER 10

Sabotage

In 2007 I received an email from a man called Steve Sosebee. He was the chairman of a charity called the Palestinian Children Relief Fund or PCRF. He asked if I would be willing to provide heart surgery for children in the West Bank and the Gaza strip in occupied Palestine. I explained that my entire training in paediatric heart surgery was a mere six-month stint as a registrar in Glasgow, and even that was about 20 years previously, so I truthfully did not feel either qualified or capable of saying yes. He then asked if I would consider operating on adults, since they were just as desperately in need of heart surgery facilities, and he explained that the PCRF, despite its name, was willing to fund and support all types of medical treatment, even if it did not involve children.

At the time, Gaza was in violent political turmoil, but the West Bank had been reasonably peaceful for a while and was, in theory at least, under the administration of a limited form of government called the Palestinian Authority. I said that I would consider going to the West Bank if I could bring with me a medical team with whom I was familiar, and Steve Sosebee readily agreed. We would go for a week in the first instance and, if the mission was a success, consider coming back at regular intervals. We fixed a week in 2008 and I set about securing the necessary leave arrangements from my own hospital and assembling the team.

At the time, Yasir Abu-Omar was a senior registrar approaching the end of his training at Papworth, and he, like me, was of Palestinian origin. It was very easy to persuade him to join the mission as second surgeon. What was surprising, however, was how easy it was to persuade the rest of the team, none of whom had any Palestinian connections. Jon Mackay, consultant anaesthetist, David Gifford, perfusionist, Steve Bryant, surgical care practitioner, Tracey Tritton, scrub nurse, and Karen Marsden, operating department practitioner, all said yes without a trace of hesitation. They were people with whom I had worked for years, and every one of them was highly capable. I also trusted them and, perhaps more importantly, really enjoyed working with them.

We started to make our travel arrangements. I had thought that, being of Palestinian origin, my arrival at an Israeli airport may raise eyebrows, and that I might be given the third degree by immigration officials in terms of questions about my intentions for travelling via Israel; so I resolved to travel through Jordan and to cross the Jordan river directly into the West Bank and, once there, head for our destination city of Ramallah. Yasir, David, Steve and Karen decided to accompany me, whereas Jon and Tracey opted to fly via Tel Aviv. We packed our surgical instruments and set off.

My group landed at Amman Airport in Jordan, stayed one night at the airport hotel and set off for the Jordan river crossing the following day. After barely a two-hour drive through the dramatic scenery of the steep mountain slopes descending towards the river valley, we were at the crossing. We negotiated the formalities at the Jordanian East Bank of the river border without difficulty, then sat on a bus that carried us through no-man's-land across the Jordan river to the West Bank, where, I must admit, I was surprised to see the immigration building with a large and unmistakably Israeli flag fluttering over it. In my utter ignorance of the practicalities of the status quo, I did not realise that the border crossing between Jordan and the West Bank was actually controlled by Israel. The Palestinian Authority had some autonomy in the West Bank, but this clearly did not extend to border crossings. So much for trying to avoid Israeli immigration officials!

I stood in a queue with our four passports clutched in my hand. When I reached the front, the Israeli officer was all smiles. She asked a few questions, checked the passport photographs against their holders and made pleasant conversation while she stamped them. I even thought for a moment that she was being more than friendly and that she was almost flirting.

'And where will you visit in Israel?' she asked as she was handing back the passports.

'Ramallah,' I said in reply.

She snatched back the passports and her expression immediately turned formal.

'Why Ramallah?' she asked.

'We're a medical team, hoping to provide some heart surgery and training,' I said.

'Sit there!' she ordered, and disappeared with our passports.

We sat and waited. Some two-and-a-half hours later she reappeared, gave us our passports and allowed us to continue our journey. We also recovered our bags and noted that every single one of them had been opened and doubtlessly closely inspected, but all our instruments were still there. We climbed into a taxi and set off for Ramallah as previously planned. Our route took us quite close to the huge separation wall that snaked across the West Bank, keeping the native Palestinians well away from the Israeli settlements built on their territory. The wall was covered with graffiti, some of which was quite artistic: the enigmatic British street artist Banksy had in fact decorated the wall with brilliant images full of his usual iconoclasm and sedition, but I do not think he was responsible for the simple and not highly artistic graffito that drew my eye: a section of the wall had 'CTRL + ALT + DEL' wittily sprayed on it in very large, pale blue letters.

In Ramallah we checked into our hotel not far from the hospital in the late evening. At four in the morning I knew without doubt that Jon and Tracey had also arrived. They are each blessed with a characteristic and very loud laugh. Their laughter reverberated through the hotel corridors as they searched for their respective rooms. They must have been in very good spirits indeed.

The following morning we went to breakfast at the hotel. There was a huge buffet with dozens of tasty Middle Eastern delicacies and we all overindulged, except for Jon Mackay who stuck rigidly to his hotel portion packet of corn flakes. We then piled into the back of the ambulance sent to carry us to the hospital.

In the hospital lobby it was impossible not to notice the large portrait of a smiling Yasser Arafat, then the Palestinian President, in battle fatigues. Next to the portrait, visitors were also greeted by a black diagram depicting a Kalashnikov AK-47 assault rifle with a red line through it and the legend: *Dear citizen, bearing firearms is prohibited inside the hospital.*

As the rest of the team panned out to familiarise themselves with the layout of the wards, operating theatres and ICU, I accompanied Dr Edie — a general and vascular surgeon who was charged with being our local medical facilitator — to the angiography viewing room. This was the room in which were kept the results of the investigations on potential heart surgery patients, especially the angiograms, those X-ray films of the coronary arteries and heart chambers which would help us plan an operation.

The room was tiny and windowless. It comprised a small wooden desk with an old desktop computer and two chairs.

It also contained what must have been over a thousand CDs of angiograms, each with a sheet of paper or more relating to an individual patient. The angiograms were piled high in cabinets, on shelves, on the floor, in and out of boxes and on the desk itself. I could not fathom a filing system. How on earth to choose patients in this situation? I discussed with Dr Edie and we agreed that, at least to begin with, we should select straightforward and relatively safe operations, and, of those, we would try to favour those patients who were desperately in need of surgery and those who had waited the longest.

Somehow, Edie appeared to have some knowledge of where such patient records could be found amid the chaos, and he selected a few dozen angiograms for us to look at. I explained that we could do only 10 or 12 operations at most in the short time we were there, but he said that we needed to select more, as some patients from our initial selection may not be readily contactable, and we should therefore have back-ups, just in case. We started to review the angiograms and the results of the investigations, a laborious process, since the desktop PC was very old, very slow and often crashed for no apparent reason.

In the meantime, word had spread that a heart surgery team was in town, and a crowd began to gather in the corridor outside the little office, with patients and their families asking to be given priority. Occasionally, they knocked on the door or, more often, they simply barged in to make their case. Despite these constant distractions, we were making slow but steady progress in identifying suitable patients, when Edie suddenly displayed a hitherto uncharacteristic excitement. He delved into a pile of angiograms, produced a CD and said, 'Here! Look at this one!'

I looked: it was a case of severe coronary disease, needing four or five bypass grafts, but with a heart muscle in good condition in a relatively young male patient in his early sixties. His only risk factor was diabetes.

'Why specifically this one?' I asked.

'Because he's in the hospital right now, so no need to go looking for him. He came in with a minor heart attack a couple of days ago, has had a lot of troublesome angina for a long time, and what's more, he happens to be the father of one of our own orthopaedic surgical trainees.'

We had found our first patient.

Several tedious hours later, we had selected a group of patients we considered both in dire need of heart surgery and suitable for our first venture into Ramallah. I met our first patient, whom I shall call Omar, and he was a delightful man. He had type 2 diabetes, but was not particularly overweight, and this was his only risk factor apart from his recent minor heart attack. He had a lot of angina and was really looking forward to life without this debilitating symptom. I promised him that we would deliver that and that the risk of surgery to his life would, in my hands, be less than 1 per cent. He agreed enthusiastically, and the team and I went back to the hotel, hungry and earnestly hoping that dinner would match breakfast for quality. It did. There were no corn flakes, so that even Jon Mackay had no choice but to venture into Middle Eastern food, and I think he was pleasantly surprised.

The following day we again piled into our ambulance-cum-taxi and went to the hospital. Omar was wheeled into the operating theatre and Jon quickly put him to sleep and inserted all the necessary monitoring lines. Yasir opened the chest and took down an artery from within to use as a bypass to the most important coronary artery, while Steve simultaneously took a vein from the leg to bypass the other four. I scrubbed up, put the patient on the heart-lung machine and carried out quintuple coronary artery grafting. The operation was one of the smoothest I had ever done: the leg vein and chest artery we used as conduits were of superb quality, the receiving coronary arteries were healthy and of a good size, and all in all the whole procedure was quick, easy, uneventful and most satisfying. A mere three hours after we started, the operation was over and Omar was being wheeled to the ICU.

The ICU in Ramallah is quite well equipped, but not exactly overstaffed, so there is a natural tendency to wake up patients early, rather than nurse them asleep on ventilators for most of the day, as is often the custom in the UK. Within two hours of the end of the operation Omar was awake in the ICU, having a cup of tea and chatting to the nurses and to his family, including his young surgeon son. I reflected that if all operations in Ramallah went as smoothly as this, we would have done a lot of good with relatively little effort by the time our mission came to an end.

That evening, back at the hotel, we all sat out on a large stone veranda with a post-prandial drink, enjoying the warm climate, listening to the raucous buzzing of the cicadas in the pine trees and generally feeling chuffed with ourselves. Two operations were scheduled for the following day.

The next morning I was awakened by the ringing hotel room telephone just after six o'clock. It was Yasir. For some reason, he'd woken up early and decided to phone the ICU to check on Omar. He was concerned that Omar's urine output was a little low in the last couple of hours. He said that it probably was nothing, but he was planning to go to the hospital to see what's what. I immediately got up, dressed and joined him in the lobby. There was no ambulance at that early hour, so we took a taxi to the hospital. Omar was sitting up in bed and feeling fine, and all his vital signs were present and correct, but his urine output had indeed dropped to less than 30 ml an hour.

As a rule of thumb, every hour a person usually pees 1 ml of urine for every kilogram of body weight, so that an 80 kg man should pass 80 ml of urine an hour. We couldn't figure out why that sudden drop had happened, but we treated him empirically with a small dose of diuretic. There was no response, and we saw only 20 ml the following hour. We tried a bigger dose. Still no response. We tried more powerful drugs. His urine output stopped completely. We did some blood tests which confirmed our fear that he was now in acute kidney failure. Neither of us had the faintest clue why that should happen. His kidney function before the operation was as good as could be expected in a diabetic 60 year old. There was not the slightest hiccup during surgery that could have upset his kidneys. His post-operative course was as smooth as silk. Why on earth had his kidneys seized up now?

Interesting as these questions may be, they were academic. The real problem was that we had a patient with no kidney function and until recovery took place a few days later, he needed to be supported by a kidney machine. We asked for one and were told that the only kidney machines in the hospital were in the renal unit, and they were used 24 hours a day. A severe shortage of such equipment in the West Bank meant the service already could not cope with the demand, that many kidney-failure patients here were left untreated, and that there was not a chance in hell that a kidney machine could be made available for our patient. I made a quick telephone call to the renal unit and confirmed this unhappy fact. While Yasir and I were pondering what to do next, the rest of the team had arrived from the hotel. Then one of the ICU nurses said, 'What about that machine that the Belgians donated?'

What about it, indeed? It turned out that a machine of some sort was given to the ICU by a Belgian charitable organisation, but it had never been used and was still in its box, kept in a hospital storage room. Nobody was quite sure what it was or how to use it. We asked for it to be brought to the ICU, and a reinforced cardboard box the size of a large fridge was duly wheeled in. We opened the box and, sure enough, it was a state-of-the-art kidney dialysis machine, but there was no instruction manual and the complex array of tubing needed to connect it to the patient was missing. None of us was a kidney specialist, but this machine was our (and the patient's) only chance.

I phoned Papworth back in the UK and spoke to the two anaesthetists who were most au fait with dialysis, and got as much information from them as I could. We asked the local renal unit staff to provide us with an array of disposable tubes, filters and dialysis solutions and they readily obliged. Then David, the perfusionist, and I went to one of the few internet-enabled computers in the hospital. We googled the machine make and serial number and laboriously surfed the web until we found a PDF document containing the instruction manual. We downloaded the document, printed it out and set to work.

We placed the machine and the tubes on the floor in the middle of the ICU and — watched by incredulous ICU staff and patients — slowly assembled it, doing our best to follow the instruction manual and using whatever bits of tube looked like they might work. David's many years of experience with heart-lung machines were invaluable in making things fit and joining the incompatible. Finally, after a few false starts, the machine worked. We connected Omar to it and almost immediately he began to feel better. Blood tests showed that the chemistry in his system was on the mend and the level of waste products normally excreted by the kidneys had started to come down.

Unfortunately, that was not the end of the story. Those very same blood tests also began to show liver malfunction. We did not believe it at first and repeated them. The next batch of tests was even worse. Omar was now developing liver failure. We did our best to treat it, to no avail. That very evening Omar's lungs also failed and he had to be put back on the breathing machine, and overnight he continued to deteriorate. He died the following morning, despite every effort and every intervention we could think of to keep him going. Our first operation in Ramallah had ended inexplicably in tragedy, and the entire team was utterly despondent.

The following day we arrived at the hospital, but with none of the high spirits that had characterised our initial foray. The previous day's patients had done very well, and so did the two that we took on that day, although they were at somewhat higher risk. By the end of the week, every patient we had operated on had also come through without incident, and we tried, without real success, to push the memory of the first disastrous case to the back of our minds.

We came to our last afternoon in Ramallah and to the last patient we would operate on during this mission. She was an overweight old woman who needed just a single coronary bypass graft. I vividly remembered seeing her at the initial assessment two days earlier. She was the archetypal buxom wife to her husband's Jack Sprat. He was a thin, sunbeaten and wiry man who had tried hard to monopolise the consultation against my best efforts to direct the questions to his wife. While I was explaining to her that her troublesome angina would be cured by a bypass operation, he was constantly muttering things like 'Why bother?', 'What's the point, at her age?', 'We don't want dangerous operations', 'Why not leave it to Allah to look after her?' and so forth. I tried to ignore his mumblings as I continued to direct my conversation at the patient herself.

I explained to her that this operation would improve the quality of her life, but not necessarily its length, as her heart disease was not life-threatening. When I finally said to her that an important part of the decision-making is that the risk of surgery includes a small chance of death of about 1 per cent, her husband's eyes suddenly lit up.

'You mean she could die as a result?' he asked.

'Yes, of course she could,' I said. 'The risk of that is small, but she should know about it before making up her mind.'

His subsequent mutterings during the rest of the consultation did an abrupt volte-face: 'Let her have the operation, then', 'Allah will provide', 'Let's take a chance on this' and so on. His transparently uxoricidal desires were most disconcerting, but, fortunately, the patient herself paid him no attention whatsoever. She asked some intelligent and relevant questions and decided that yes, it would be worth the risk for her to have the operation and she would go ahead.

As I walked towards the operating theatre, I bumped into a very agitated and anxious-looking Jon Mackay just outside the door to the anaesthetic room. I asked him if he was all right.

'Never mind. Just get in there and do the case,' he said,

'and I'll tell you about it later.'

About what? I insisted that he tell me there and then. He said that something very strange had just happened. He went on to explain that one of the differences between Ramallah and what he was used to in Papworth was the absence of a capnograph. This is a device that confirms that a patient on a ventilator is receiving oxygen and exhaling carbon dioxide. It is a sure-fire way to check that everything in the ventilator and its tubing circuit is in working order. Ramallah had no such machine, so Jon was exceptionally meticulous in checking the ventilator circuit. On that occasion, he and Karen Marston had prepared the anaesthetic drugs and double-checked the circuit carefully. Mrs Sprat's obesity meant that she could be difficult to intubate for ventilation and so Karen had, in addition, deployed a bougie to help ensure the tube went into the right place, which is the windpipe.

There was a delay sending for the patient, so he and Karen had retired to the coffee room with the local team. On their return, the bougie had gone 'missing'. This was disappointing but not life-threatening as we had a spare available. However, despite having already checked the ventilator circuit, Jon, displaying some obsessive-compulsive behaviour, decided to check it again and was alarmed to find it had been compromised. There was a disconnection leading to a leak of anaesthetic gases, which he eventually traced and located. In 20 years of clinical practice, he had never encountered such an event in a recently checked system. The location of the anaesthetic machine in the theatre and the location of the disconnection made it extremely difficult to believe that the disconnection was accidental. Moreover, it required specialist anaesthetic knowledge. Had Jon not obsessively repeated his checks, Mrs Sprat would have died from oxygen starvation soon after being put to sleep. It was this discovery of evidence of tampering with the circuit that had made him extremely nervous, and he believed that deliberate sabotage was a distinct possibility. Mrs Sprat was already in the operating theatre and the most immediate and difficult question was 'Do we continue?'

We decided to continue. Mrs Sprat had an uneventful operation and all went well. When the operation was over, my thoughts went back to our first patient, Omar. As it happened, the rest of the team were having similar concerns and, that evening, Steve was the first to voice them. Given the unexplained and relatively late onset of sudden and fatal failure of Omar's kidneys, liver and lungs a few days earlier, we had all started to think the unthinkable. Our last case was clearly a near-miss victim of deliberate sabotage. Was our first case equally sabotaged? Could Omar have been intentionally poisoned? This was the Middle East, and Islamic tradition requires burial within 24 hours of demise, so the possibility of a post-mortem examination was out of the question.

We simply did not know. When Omar died, we had put it down to a freak event we were unable to understand, or some complication of a systemic disease that we simply did not know he had, or just terribly bad luck. But the events around Mrs Sprat were far more worrying, because Jon had direct evidence that someone must have interfered with the ventilator circuit in the short time between his careful setting up and the start of the operation. Only our first and last patients were involved, one in a calamity that actually happened and the other in a calamity that we had only just narrowly avoided through Jon's fussiness and perspicacity.

That night we were all invited to dinner by the Minister of Health to thank us for our efforts. After much deliberation, I decided tentatively to approach him with our suspicions. He did not seem to be unduly surprised, but did promise to investigate the matter fully and send us a detailed report of the investigations. Nine years have now passed and I have still not seen a report.

On our return across the bridge to Jordan, I was accosted by an Israeli man in uniform who declared that he was from the 'Ministry of Tourism'. He started by asking questions about our stay that could be related to tourism, and I answered these. Then the questions deviated away from tourism into the names and addresses of my family members, friends and acquaintances, and I became suspicious. It was apparent that he was not at all interested in tourism and I refused to answer any further questions.

During that time, Jon Mackay and Tracey Tritton were on their way to Tel Aviv Airport. I had hoped that at least they would sail through the formalities easily. I was wrong. They'd had a hard time, too, and Tracey was quite reticent about it until, many years later, when I started writing this book. I asked her to let me know what happened and also asked her and other members of the expedition to check over my story above to ensure I had indeed got all the details right. She more than obliged: she wrote it all down and, having read her account, I feel that there is little to do other than reproduce it in its entirety and using her own words:

> After a week away from my husband and daughter, I am finally going home. I never expected and was completely unprepared for what awaited me as I departed through immigration.
>
> This was my first visit to Palestine. I was very moved by the political situation, in particular the impact of the Israeli occupation of Palestine and what it means to be a citizen of the West Bank.
>
> My experience with border security throughout my visit and upon leaving Israel by plane was beyond surprising. On the drive to the airport, I felt that every one of the many checkpoints would be the one to stop us from actually making it. The second checkpoint was the worst. Jon and I were asked to get out of the taxi and lift our cases from the boot. We were asked to open the cases. One armed guard disappeared after taking our passports; the other armed guard watched us both unzip our suitcases. I felt physically sick. Here I was on my knees opening my suitcase with an armed guard stood behind me.
>
> The other guard returned with our passports and

said we could go. I remember sitting with Jon in the back of the taxi, thinking, 'What if we never get to the airport? What if we never make it home?' I also recall that my legs wouldn't stop shaking with fear.

I had witnessed and experienced first-hand the daily oppression of Palestinians, yet still I was unprepared for the direct experience with Israeli security that I met at the airport.

Jon and I arrived at the airport four hours early, because travellers are told to allow three to five hours for security. We had walked what seemed like only a few yards through the terminal doors when a security guard approached us both and asked to see our passports. She asked where we had travelled from and what the nature of our visit was. We said we had been asked by a humanitarian charity to carry out some medical training and heart surgery. When asked where, I replied 'Ramallah'. She wanted to know who originated this request and if we had any further documentation to prove this. I produced the original invitation from the Palestinian Health Minister, and the guard then walked away with both our passports and my invitation letter. Within minutes she was back and told us we could proceed to the scanning of our bags area. I honestly thought that was it; not too bad if that's the worst questioning we get. The guard and three of her colleagues kept glancing over. Jon was calm, but I could sense he was worried. Jon normally has a wicked sense of humour, so when he suddenly turns serious and quiet I know this is a time to worry.

There was an unbelievably long queue immediately before us. Before even approaching the check-in counter we had to have our luggage scanned, but this didn't bother me as it is now common practice in many international airports. Our luggage was scanned and then I was ushered to a nearby table for a 'random-selection search', or so I was told. It took a substantial amount of time as the search was meticulously detailed and conducted in the open, just a few feet from the entrance. Every single item of my clothing was taken out and put back. One item they did find of interest in my case was a plaque I had received as a token of thanks from the Palestinian charity and health minister. They kept passing it to each other and laughing, and it appeared as though they were mocking me. It was horrible.

An hour had passed since our arrival and we had yet to check in for our flight. Jon also had his case looked through, but not to the extent that mine was examined. They said he was free to go.

It was not the same for me.

I felt the presence of two guards behind me, and I was asked to go with them for questioning. My case remained open. I asked Jon to keep an eye on what might happen to it. I was petrified they would plant something in the case to prevent me getting on the plane.

As I was escorted to an area on the other side of

the airport by the two female guards, I began to cry. I asked why I was being taken away. One guard appeared humane, but the other was what I could only describe as brutal.

I was taken into a room. One guard took my rucksack and passport and the other guard told me to take off my clothes. I asked if I could keep my underwear on. This they refused. Once I was undressed they ran a scanner over all parts of my body. I felt vulnerable and childlike. I could feel tears running down my face, and kept having flashbacks of images of both my husband and daughter. What if I never got home? I kept running over in my head every possible scenario that might occur, and I realised I had no control over my destiny.

They took my backpack and emptied the entire contents on the floor. They looked at the photos in my wallet, while laughing and chatting among themselves. It was horrible and unnecessary.

The guards left me standing naked in a side room of the interview area, and eventually told me to sit on the chair next to a table. One guard sat opposite me, the other stood behind me with her rifle.

I asked if I could get dressed. The armed guard behind me informed me that I could get dressed once they were satisfied with my answers during the interview.

The guard sat down and the following exchange took place:

Guard: 'What was the purpose of your visit?'

Me: 'We were asked by a charity to help train some staff at a hospital in West Bank and carry out some cardiac surgery.'

'Which hospital?'

'Ramallah.'

'Which charity?'

'PCRF, Steve Sosebee and the Palestinian Health Ministry.'

'Why Palestine?'

'I would have travelled to Israel if the invitation was from Israel. I don't support one side more than the other. I am here to simply do my job as a theatre nurse.'

'Who else was in your team?'

'Jon Mackay, Sam Nashef, Yasir Abu-Omar, David Gifford, Karen Marsden and Steve Bryant.'

'Where are the rest of the team?'

'Still in Ramallah.'

'Will they be travelling later back through Tel Aviv?'

'Yes.' (I knew this was not true, but I didn't want to cause problems for the rest of the team getting out across the river to Jordan.)

The questions continued for over an hour and a quarter. My legs continued to shake. I remember sobbing, with tears running down my face on to my bare body. I kept asking please could I at least wear my top. The guard with the gun shouted at me and told me to stop crying and be quiet. I think it was at this point that I lost control of my bladder. My dignity was totally destroyed. I felt humiliated. Both guards were laughing at me and speaking to each other in Hebrew. I desperately wanted to get out and go home.

Eventually, the guard who sat opposite me said they were satisfied with the questioning and I could get dressed. They would escort me back to my case and to passport control. This would give me only 15 minutes to get through passport control and board the plane.

I continued to sob. The armed guard threw my clothes at me and said, 'Be quiet! Get dressed!'

The other guard tried to comfort me by saying that what I have just gone through is routine and normal.

I continued to cry and asked them why they were doing this and they both looked at me and laughed. I was then escorted back into the airport complex.

At passport control I was greeted by Jon. He gave me a huge hug and tried to make light of the situation by saying, 'You should have told them you'd killed a Palestinian, Tritton!'

Typical Jon humour, but all I could do was sob and think that it wasn't over until I was on that plane and it had taken off, and I knew for sure I would be safe and on my way home.

I was ignorant of the details of what had happened to Tracey and remained so for many years until I actually read the account above. I was, nevertheless, seriously perturbed by the events that had already transpired, and so, in our last few hours in Ramallah, I had taken the opportunity to ask questions. I found out that similar ventilator 'problems' had afflicted and occasionally killed heart patients operated by visiting surgeons in the past. I was told that there was quite a bit of political and professional rivalry in the medical world around Ramallah, and that it was not in some people's interest for a heart surgery programme to succeed there. I also found out that the Israeli intelligence services are known to recruit collaborators in the West Bank to do their dirty work, and the Israeli authorities had made it abundantly clear that they were not too keen on our mission, but would any of these people stoop so low as to risk killing patients?

To this date I truly do not know the answer to that question. Recently, Amir-Reza Hosseinpour (a superbly skilled paediatric cardiac surgeon from Seville in Spain, who had trained at Papworth in the past) led a mission to Ramallah to start a paediatric cardiac surgery programme there, along the same lines as ours, but aimed at children. Every operation he carried out went without a hitch.

Meanwhile, Jon Mackay read Tracey's account of the airport shenanigans and, sense of humour still intact, suggested that this book be entitled *The Naked Scrub Nurse*. Although Tracey laughed at the tasteless joke, neither she nor I were very keen on the idea.

CHAPTER 11

Irony

One of my contributions to the field of medicine is the EuroSCORE model. This is a system which takes a few characteristic features of a heart patient and calculates the likely risk of death or disaster from a heart operation on that patient. Knowing the likely risk of an operation helps the patient and the surgeon decide whether or not it is worthwhile to go ahead. It also helps in measuring the quality of a heart surgery programme. By comparing the expected risk of death in a group of patients with the actual death rate when the operations are carried out, we can form an idea about whether the unit is performing as expected, better than expected, or worse.

Yesterday I received a phone call from Swetta, one of our surgical trainees. She was regaling me with a number of unexpected and highly unpleasant post-operative complications that one of my patients had developed all at once.

She wanted my opinion and advice on how best to manage these complications. Once we had finished the conversation and had formulated a plan for the patient, she declared, 'By the way, I think this EuroSCORE system of yours is a bit rubbish, because there is an important risk factor missing in the calculations.'

'Oh really?' I asked. 'What risk factor would that be, then?'

'It's simple,' she said. 'The patient being a doctor should be considered a EuroSCORE risk factor.'

She was speaking tongue-in-cheek, of course, but alluding to the fact that the patient in question happened to be a doctor: he worked as a general practitioner nearby. On top of that, many of his family members were also doctors and his son was a cardiothoracic anaesthetist, working within my field of cardiac surgery in a unit in the north of the country. If ever one wanted an operation to go smoothly … and yet here he was, having most of the complications a patient could have after a heart operation all at once. The trainee was expressing the commonly held view that doctors make the worst patients, and by that she did not mean that they can be demanding. She was referring to the widely held perception amongst all of us that it is always the bloody doctors who get the worst complications and do badly.

The irony of such happenings is not lost on us surgeons. I do not think that there is any statistical truth in the assertion that doctors do less well than other patients when they have operations, but despite that, I cannot help approaching any patients who themselves are medically qualified with some trepidation, always expecting the worst. Even more so, the trepidation increases the closer I feel such patients are to my own profession of surgery; and it also increases in direct proportion with their perceived importance in the medical world, and with their geographical proximity to my hospital, Papworth. I guess that the maximal possible stress will occur when I am referred a fellow cardiac surgeon at Papworth who is also very highly prominent internationally. This has not happened yet, but it may only be a matter of time. In the meantime, I sincerely hope (for purely selfish reasons) that all my colleagues will continue to have happy and healthy hearts for the rest of their lives, or at least until I retire from active practice.

For over 20 years I have run a programme of hybrid surgery for coronary disease. The overwhelming majority of my patients needing treatment for angina get the full works of a CABG, with long incisions in the chest and leg. A few, like Emma in Chapter 6, had only one diseased vessel and could have a minimally invasive CABG, with a tiny incision in the chest. Another small group of angina patients have blockages in two vessels: one that can be dealt with by minimally invasive CABG, and another that is suitable for a stent. It seemed to make sense to treat such patients with a hybrid approach, so that I would do a minimally invasive CABG for one artery and then, a couple of days later, a cardiologist would stent the other artery and, hey presto, the coronary disease is sorted in its entirety without splitting the breastbone. I teamed up with

Peter Schofield, a trusted Papworth cardiologist and friend, to offer this service and, over the years, we have had fantastic results in about 100 patients treated this way and we can boast of having possibly the world's largest experience in this unusual method. It is, of course, a very attractive approach for patients in that it avoids big cuts in the chest and because recovery from it is relatively quick. The other feature of this approach is that, two days after my bypass graft, Peter would have to do an angiogram to prepare for implanting the stent, and that angiogram would of course include taking pictures of my recently done bypass graft, a harsh and unforgiving system of quality control to which most bypass grafts are not usually subjected; and I used to enjoy crowing about the 100 per cent success rate by having all the bypasses in this group checked by angiograms and being able to prove with hard X-ray evidence that all my bypasses were working fine.

When Sion Lewis, a man in his early sixties, was referred to me for this specific procedure, he fitted the bill perfectly and it appeared that he was most definitely a suitable candidate for this hybrid approach. Looking a little more closely at the referral, some trepidation began to surface when I realised that Sion was himself a doctor. The trepidation rose sharply when it turned out that he was an orthopaedic consultant surgeon, and rose still further when I found out that he worked in a hospital a mere 15 miles away. By the time I discovered that he was also the son of Ivor Lewis — the famous thoracic surgeon who is recognised worldwide as the brilliant originator of the Ivor Lewis operation for cancer of the gullet — I was truly worried.

When I saw him in clinic, Sion told me that he had opted for the hybrid approach in the hope of completing his post-operative recovery in the shortest time that could be achieved. This was so that he would take the minimum time possible off work. In fact, we planned his bypass on Tuesday, his stent on Thursday or Friday, and he was to go home by the weekend, so that he could be back at work by the following Monday, when he himself had a full operating list. It all seemed a bit ambitious, but I remembered another of my patients, also a consultant orthopaedic surgeon, who was happily hammering in new hips only three days after this type of operation. So I assured Sion that yes, what we had envisaged was actually possible, provided everything did go according to plan. Life, however, is what happens to you while you are busy making plans.

We went ahead with the minimally invasive bypass on Tuesday. Three days later, instead of having his stent done, Sion was otherwise preoccupied in nearly dying from the most virulent wound infection I had ever seen in my career. It had spread to his bloodstream, made him feverish, delirious and so sick that he had to be readmitted to the ICU. No amount of powerful intravenous antibiotics touched it, and I had no choice but to take him back to the operating theatre to clean and wash out his septic wound under general anaesthetic. It took him a good six weeks before the infection finally settled. And the bypass graft? When Peter did his angiogram it was an utterly useless, blocked bypass.

Fortunately, Peter's stent was more successful. Sion was cured of angina and is now back at work, but not a scintilla of this good result could be ascribed to anything I did for him. All I had achieved was to give him a life-threatening wound infection and a useless graft. Oh, and my success rate for these procedures has dropped down to 99 per cent, and I crow a lot less about them nowadays.

The cruel irony that seems to drive surgical complications is not exclusively confined to doctors as patients. It is something of an open secret that I compile cryptic crosswords for two national newspapers, *The Guardian* and the *Financial Times*. I find this a most enjoyable hobby which indulges my love of language and wordplay, and provides an excellent distraction from the sometimes harrowing pressures of the day job. Both of these newspapers provide their puzzles on line for free, so that at midnight, well before the newsagents have opened their doors to the morning commuters, the next day's crossword is already available for keen solvers on each of the newspapers' websites. As a result of this, a small community of online bloggers has formed around a number of websites, and there is a bit of a race to be the first to declare completion of a puzzle. The fastest solvers do so in 20 minutes or thereabouts.

These online blogs, as well as serving as a means of communication for the crossword-solving aficionados, also provide invaluable feedback to the crossword compilers. Many of us look at them to see how the bloggers react to our puzzles. Too hard? Too easy? Fair or unfair clues? Any errors or howlers? Witty or boring? The bloggers publish their verdicts on the websites, and these verdicts can sometimes be harsh. One of the websites is called *Fifteensquared* and has the amusing motto 'never knowingly undersolved'. It features several bloggers, one of whom used to blog under the name of R. C. Whiting, who came across as a hyperintelligent and widely educated person not easy to please with a crossword puzzle. Whiting blogged freely and critically whenever he deemed a clue unsatisfactory. As a general rule, I appreciate all the feedback that my puzzles receive on *Fifteensquared* and usually read the comments as soon as they appear. I know that some other compilers do the same, and some of them also respond and engage with the comments, whereas my friend the late great John Graham, better known as Araucaria, the doyen of crossword compiling, never once looked at the blogs.

The next new patient to enter the consulting room at my usual Wednesday afternoon clinic was a middle-aged man. When he sat down, he casually placed on the edge of the desk a copy of *The Guardian* newspaper, folded to show a partially completed cryptic crossword puzzle, clearly something he had brought with him to help pass the time in the waiting room. I introduced myself, shook his hand and, having spotted this potential interest that we had in common, I asked him how he was getting on with the crossword.

'All right,' he said, 'but it's not my favourite compiler.'

I asked him who was his favourite and he replied without hesitation 'Arachne', because of her precision and her wit. I agreed with him wholeheartedly about Arachne, who is also a favourite of mine, and he then announced that he was very active on the *Fifteensquared* website.

'How very interesting,' I said, 'under what name?', and he looked at me with a bemused expression and pointed to the name clearly written at the top of his own medical notes: Roger C. Whiting.

When we finished talking about crosswords, we discussed his angina, which was indeed troublesome, in that it occurred on exercise, but also sometimes bothered him at rest. He was keen to get rid of it. We agreed that a bypass operation would be a good idea. He had been a smoker and also had some risk factors which made his operation somewhat riskier than usual, but there was certainly nothing to prohibit going ahead. I quoted him a risk to life and a risk of stroke of around 1 to 2 per cent and he was more than happy to proceed on that basis.

This was at a time when we were so busy at Papworth we had no space for all the operations we had to do, and were forced to use a private hospital in London to carry out some extra operations, so as to help us cope with the demand and to reduce waiting times for patients. I booked Mr Whiting for surgery in London and, a few weeks later, did a perfectly straightforward CABG operation with a quadruple coronary bypass.

A few hours later I was horrified when he woke up with a stroke. What's worse, because of some features in the manifestation of the stroke, I initially misdiagnosed it as Parkinson's disease and gave him treatment which, for a short while anyway, actually made him worse. His recovery was slow and laborious. I brought him back to Papworth soon afterwards and he only managed to leave hospital to go to the stroke rehabilitation unit a month later and, for several weeks afterwards, the mere prospect of solving a cryptic crossword was out of the question for him. When he was discharged from hospital I gave him one of my unpublished puzzles to practise on, and if you fancy pitting your wits against Mr Whiting's, the puzzle is at the end of this book.

Some weeks later, I was delighted to receive the puzzle in the post, completed correctly, albeit in somewhat shaky handwriting. The stroke had affected his ability to write and it was a quite a while before Mr Whiting was again able to blog on *Fifteensquared*. He eventually started again, having made a full recovery, and whenever he said something nice about one of my puzzles, I felt very grateful — and in more ways than one.

Finally, irony strikes at the heart of Cambridge science. The veteran Cambridge cardiologist Michael Petch referred a patient to me who was both a local and an international celebrity. Professor Max Ferdinand Perutz (1914–2002) was the Austrian-born but Cambridge-based molecular biologist who had, in 1962, won the Nobel Prize for Chemistry for discovering the chemical structure of haemoglobin, the protein that makes up blood, carries oxygen to the tissues and is the reason why blood is red. In short, Max Perutz discovered the secret of blood. His other massive achievement at Cambridge was that he had founded and chaired the Laboratory of Molecular Biology, 14 of whose scientists went on to win further Nobel Prizes. Amongst them were James Watson and Francis Crick, who were part of the team which discovered the structure of DNA. Max Perutz was in his early eighties and had angina when Michael Petch referred him to me, and despite his age Max was intellectually as sharp as ever. He just needed a single coronary bypass, and Michael left the notes on my desk with the specific instruction that I should *Look after his brain – it's one of Cambridge's most valuable assets!*

So I did my best to look after his brain. I did the entire operation myself, skin to skin, for speed. I was careful to keep his blood pressure high throughout the time under anaesthetic and while he was on the heart-lung machine. I was as time-efficient as I could be, using the heart-lung machine for only 15 minutes, and at the end of the operation I felt as confident as I could possibly be that his brain would have come through intact. Two hours after Max's return from the operating theatre to the ICU, I was called urgently to review him. I walked to the ICU with an awful feeling that there was some dire problem with his head when he was woken up. There was not. There was, however, an awful lot of blood in his chest drain bottle, and it was still coming. He was bleeding profusely. There was nothing for it but a return to the operating theatre to control the bleeding. I reopened his chest and found the culprit: a tiny side branch of the internal mammary artery which I had used for the bypass was spurting away: a little red fountain which took a couple of seconds to clip shut. Max Perutz's heart and his brain survived intact and he did very well over many years until he sadly died of something else.

I have previously mentioned the McKlusky's Club, where heart surgeons of my generation present our disasters, so that we can learn from each other's mistakes. That year my talk at McKlusky's was the shortest I have ever given, consisting of only three slides and a single sentence:

> **This** (Slide 1) is Max Perutz, who won the Nobel Prize for discovering **this** (Slide 2), the structure of haemoglobin, and **this** (Slide 3) is a chest drain bottle full to the brim with Max Perutz's own haemoglobin after yours truly operated on him.
>
> Thank you for your attention.

CHAPTER 12

McKlusky's

The establishment of 'generation' clubs has been a tradition in British cardiothoracic surgery for several decades. What happens is that a group of consultants in the specialty get together and invite all other consultants from their vintage to join the club. Vintage in this context is determined by the year in which a surgeon was appointed to a consultant position. Over a few years, the membership slowly expands as more surgeons join, until, when the club reaches a sufficiently large yet manageable size, it is closed to new members and a new club is established by the next generation.

Most of these clubs meet once a year and, in the past, their meetings used to be lavishly funded by medical device manufacturers in the commercial sector. Over the years, the authorities which regulate doctors' behaviour have increasingly frowned on such sponsorship, fearing that doctors may be unduly influenced and biased towards using products made by manufacturers who provide them with such hospitality. As a consequence, such activities have gradually become more strictly regulated and most of the funding has dried up. Nowadays, the meetings are financially supported and paid for by the members themselves.

The club to which I belong is called McKlusky's and is named after Dan McKlusky's Steak House, an unprepossessing small restaurant in Austin, Texas, where, a quarter of a century ago, the club's founders first met to discuss establishing such a club. They were British delegates attending an American surgical meeting at the time and the reason for their presence at McKlusky's was simply that they were hungry and needed to eat. The restaurant, by all accounts a modest one, has long since closed its doors, but the club survives to this day. Other clubs are named after much more fancy eating establishments.

We meet once a year at a weekend in mid-January and take turns to host the event in the cities where we work. We have a scientific day in which we present cases and a social day for a bit of fun and, occasionally, some culture. One feature that distinguishes McKlusky's Club from other surgical gatherings is the nature of what is presented and discussed at the meetings. Most surgical professional meetings are filled with either scientific research or with surgeons reporting, in one way or another, a series of operations with phenomenally good results. In other words, the surgeon stands on the podium and delivers a lecture that invariably has the underlying subtext: 'Look how clever I am.'

Not so at McKlusky meetings, where members are invited to present only those cases which reflect badly on them: where things went wrong, mistakes were made and poor outcomes were the result. The idea behind this is a sound one. It is far better to learn from other people's mistakes than from one's own, and sharing this knowledge will draw the group's attention to an unexpected or unknown peril of heart surgery, allow discussion of what could have been done differently and send the members home having learnt to avoid the pitfalls that have trapped their colleagues. A prize is awarded for the greatest disaster, and the unhappy winner's name is inscribed on a plaque, which is taken home for a year until it can be handed to the next unfortunate 'winner' at the following meeting.

I regret to say that during the quarter century of annual McKlusky's meetings, I have the dubious honour of having 'won' the prize on no fewer than three occasions, for reporting disastrous cases, some of which are retold in this book. At the last meeting of the club in Wimbledon I asked my fellow members to write down the most poignant disasters that they had presented to the club over the years. I promised them anonymity and, of course, patient confidentiality, but I wanted the stories told in their own words. Two of them responded (I shall call them 'H' and 'J') and set down in their own words a pair of poignant tales which I still remember vividly from the day they were presented to fellow members of the club. Here are their stories.

THE STORY FROM SURGEON H

The newborn baby had a straightforward 'hole in the heart'. Or at least it should have been straightforward. But there was a clue. Referrals for heart surgery are not meant to come from the children's lung doctors (respiratory paediatricians) — they usually come from the cardiologists (heart doctors).

Children's heart surgery is stressful. Congenital heart disease is uncommon, but the variety of possible defects is vast. So, as a surgeon, most of the operations you only do a few times per year. You don't have the chance to get into your comfort zone. Nights of disturbed sleep before the operation, going over each step, are the norm.

But this was a ventricular septal defect (VSD): a hole between the two pumping chambers (ventricles) of the heart. The commonest congenital heart defect requiring open heart surgery. So this was a pleasant distraction. Closure requires a patch. Even though a baby's heart is about the size of a plum, it is a standard operation (we wear magnifying glasses).

The physiology of a VSD is fascinating. At birth, the lungs are still quite stiff with a high resistance to blood flow. In fact, while the baby is still in the womb, hardly any blood at all flows through the lungs. As a result, when the baby is first born, the pressure needed to pump blood through them is high — the same as that in the other side of the heart (which is pumping blood around the body). So even if there is a hole between the pumping chambers, not much blood flows through the hole, since the pressures are equal. Over the first two to three months of life, the lungs naturally relax and the blood pressure in them falls to about a fifth of the body blood pressure. As it gets easier for blood to flow through them, more and more blood goes through the hole, eventually flooding the lungs with excessive flow. So an operation to close the hole is typically performed when the baby is about three months old.

This baby was different. On a ventilator (breathing machine) from birth, his lungs were a problem — a big problem. The intensive care team could not get the baby off the ventilator. The respiratory paediatricians pleaded with me to close the hole. I explained the physiology: the VSD was not causing a problem (yet). They argued that this was not a baby with normal lungs. They said that they would be happier if the hole in the heart was closed. Then they could focus on the lungs. I relented.

I met with the parents and explained the situation. Closing the VSD was a low-risk operation, I heard myself saying. It would probably not make things better — the lungs were the problem. But it would take the VSD out of the equation and the team could focus on trying to work out the problem with the lungs.

The baby was to be the second case on the list, and the operation usually takes about four hours. I told my wife I would definitely be home for dinner. And told the kids I would see them before they went to bed.

When you open the chest (through the breastbone) to get at the heart, it is not uncommon to open one of the two pleurae (the cling-film-like lining around each lung) as they meet in the middle. It is not a problem — usually. The left pleura was opened and the left lung immediately protruded out of the chest (very unusual). But I got on with the operation. I connected the baby to the heart-lung machine. I stopped the heart. I opened the right atrium (the collecting chamber on the right), so that I could see the hole through the tricuspid valve (between the right atrium and the right ventricle). I sewed the patch in place. Closed the heart. The anaes-thetist connected the ventilator. All very routine.

But the left lung looked ominously over-inflated. It would not go back in the chest. We weaned the baby off the heart-lung machine: the perfusionist gradually reduced the flow of the bypass machine and this allowed the heart to take over the circulation. But the strain on the heart was too much. The heart was unable to fill with blood, because of the pressure from the very abnormal, over-inflated lungs.

We went back on the bypass machine and rested both the heart and the lungs. We tried again to wean: same problem arose. We waited. We watched. We called the cardiologist. We racked our brains. We tried again. By this time it was late evening. In the end we had to admit defeat. The baby was not going to survive. We stopped the machine and the anaesthetist turned off the ventilator — the lungs relaxed. I left my assistant to close the chest and I went to see the parents, who were waiting anxiously in the ward for news.

Taking the senior nurse with me, I sat down with the family in the Sister's office. They anticipated bad news, because I was so late coming to see them. I explained that their baby was dead — the hardest thing you could ever have to do. They were upset, but not surprised. There were tears. Then silence. I left to go back to theatre to help the team and to do my notes.

As I walked along the corridor, one of the theatre nurses was running towards me.

'Have you told them?' she said.

'Yes.'

'Well, the heart is beating — you need to come and see.'

In trying to get the chest closed, my assistant had opened the other pleura to take the pressure off. And the heart had started to beat! So the anaesthetist had reconnected the ventilator. Even though both lungs were sticking up out of the chest, the heart was beating and the blood pressure was good. So we watched, thinking it would all peter out. It didn't. Things seemed remarkably stable. We were not able to close the chest. In the end we covered the protruding lungs with a dressing and took the baby back to the ICU (intensive care unit).

Then I had to go and see the parents again. What

> could I possibly say? How could I explain this to them? I can't remember what I said. What they felt when I told them I had been wrong, that their baby was not dead, I cannot imagine. But they were surprised, delighted, shocked and thanked me!
>
> I learned an important lesson: while communication may be vitally important, timing is everything.
>
> (written in H's own words)

Listening to my friend and colleague telling us this harrowing story, I was immediately reminded of the time I was a house officer — the most junior doctor on the medical service — and had to deal with a similar event, but at the other end of the age scale.

I was on call at the weekend and covering a number of wards when I was paged to see a Mr Brown on the geriatric ward. The nurse who called me explained that he had died quietly, that he had been expected to die soon anyway and that the medical team looking after him had no plans for any dramatic resuscitation should it happen. The only reason she called was that he had to be certified dead by a doctor before the mortuary could be contacted.

I had ambled at a leisurely pace towards Mr Brown's ward and walked straight to the one bed with curtains drawn around it. Mr Brown was clearly an old man well into his eighties and he was sitting motionless in a standard NHS armchair by the side of his bed. I felt for the pulse in his neck and thought that I could detect something weak and thready there, meaning his heart might still be beating, and then I noticed that despite this he was not breathing. I immediately moved his jaw forward to open up his airway and he instantly took a deep breath, and opened his eyes. He most certainly was not dead yet. I helped him back into his bed, drew open the curtain and went to tell the nurse in charge the good news. She was walking out of the nurses' office, saying, 'Thanks for that, I've already told the family. They took it well — they were expecting it.'

Whoops.

We had to do something immediately. After an initial panic, we phoned the family back, lied through our teeth, put on some frantic theatre act and pretended that he was being resuscitated, and that we'd call again with the outcome. Then we called again a few minutes later to report the successful outcome of the non-existent 'resuscitation'. Whether the family were relieved or upset to hear the news, I was never quite sure. Nowadays, 'do not resuscitate' orders must be discussed with the patient and family before being instituted. We all follow a duty of candour, of being totally honest with our patients, even if the truth hurts, and we would have never got away with what the nurse and I did so many years ago.

THE STORY FROM SURGEON J

'Looking out for my Frank'
a personal perspective of reflection over
a lost opportunity

'To err is human …'

Alexander Pope

'Every surgeon carries about him a little cemetery, in which from time to time he goes to pray, a cemetery of bitterness and regret, of which he seeks the reason for certain of his failures.'

French surgeon René Leriche (1879–1955)

'If you can meet with triumph and disaster …'

Rudyard Kipling

These quotes are familiar to many experienced surgeons and describe the mental anguish which is inevitably associated with surgical failure.

Successful surgery is about choosing well, cutting well, and getting well — but it's also about resources. Frank was a 67-year-old man with a serious condition who should have had a relatively low-risk heart operation with a successful outcome. He had worsening chest pain and dizziness for 12 months and had seen a cardiologist in a district hospital. The cardiologist made the diagnosis of a narrowed aortic valve and referred him to our hospital, which provides both standard heart surgery and transplantation. Tests confirmed the narrowed valve and showed some coronary artery disease, which also required attention: one of his coronary arteries had a tight narrowing in it.

When I saw Frank in the clinic, he was already feeling worse and had breathlessness on exertion and swollen ankles. His breathlessness was bad enough to prevent him having a good night's sleep. This is a sign of the heart beginning to fail and of fluid build-up in the lungs as a result, which makes the situation urgent: the heart cannot cope with the narrowed valve for much longer and sudden death becomes a real possibility. I wrote that *He should be admitted in the next four weeks.*

The operation was planned accordingly, but at the last minute it had to be postponed, because of a more urgent case being transferred to us from another hospital. We rescheduled the operation for the first available space, 11 days afterwards, on the same operating list as another urgent patient.

The night before, a heart transplant operation was carried out. Heart transplants occur out of the blue when a donor heart becomes available, and they are often carried out late at night or in the small hours. When that happens, substantial numbers of hospital staff are up all night and it can affect the personnel level available the following day. We did not have the staff to run the operating theatre for either operation the following day, so both operations had to be cancelled. The waiting-list nurse gave him the bad news, told him to pack up, go home and wait for the next operating slot.

Frank, clearly very despondent, got dressed and waited in the ward day room for his family to come and take him home. At about the same time as the family arrived on the ward, he had a cardiac arrest. His aortic valve narrowing was severe and his heart was teetering on the brink of failure. So, predictably, all attempts at resuscitating him from this were unsuccessful. He died from a condition that was eminently treatable, because the treatment was not given in good time.

The NHS is ridden with delays. It took six weeks after Frank's family doctor referred him to the district hospital for him to be finally seen in the cardiology clinic. It took another two weeks for the tests to be done: the echocardiogram to look at the valve and the angiogram to look at the coronary arteries. It took another five weeks before the referral letter to the surgical service was received. It took another four weeks before he could be seen in the surgical clinic. Once that letter was received, we saw him within a week and scheduled his operation for two weeks hence, and then the two cancellations happened. Even so, five months had passed between him being referred to hospital and the first operation date. This is too long a delay for a patient precariously balanced on a knife edge.

It later transpired that Frank, over five years, had generously shared his heart murmur with a succession of medical students, and was told of the clinical urgency. There was delay in communicating the additional finding of coronary artery disease, and repeated family phone calls, generally received by an answer phone, were met with impatience.

Severe aortic stenosis follows a predictable downhill course unless surgically corrected and there was an undoubted accumulation of multiple critical delays which resulted in a catastrophic outcome for this patient and his family.

Referrals for complex surgical procedures nowadays tend to be process-orientated and there is a lot more 'disconnect' compared with when I first began as a consultant, when there was just me and a secretary. But, of course, so much has changed: accountability, a more complex case mix, and, most importantly, a significant increase in patient demand, which has not been met with a corresponding increase in resources.

Could I have tried harder, as the lead advocate for this patient? Everyone who works in the health service knows that there are so many competing interests for theatre space. At times our transplant activity and our endless battles to get cases done with theatre staff, surgical colleagues, anaesthetists and intensivists feels like wading through wet concrete. Perhaps I was insufficiently attentive, and the patient paid the price.

The words uttered by his wife will forever remain with me: 'Nobody was looking out for my Frank.'

(written in J's own words)

When I first started as a consultant heart surgeon in 1992 I became concerned by the speed at which my waiting list was growing, so that, after a relatively short time in the job, some of my patients were already having to wait well over a year for surgery. Every now and then, one would die from the heart disease which the operation was intended to fix. I was not alone in this: deaths also happened on all of my colleagues' waiting lists. The obvious conclusion from all of this is that waiting for a heart operation is risky. This is not altogether surprising: many of our operations are done at least in part to prolong life, so it stands to reason that not doing the operations in a timely fashion would put the patients' lives at risk.

I decided to study the risk of waiting for a CABG and collated all the data from our hospital's waiting-list office. The result was a short paper published in 1996 in *The Lancet* medical journal. The findings were appalling: the rate of death on the CABG waiting lists overall was 1.65 per cent, which is about the same as the rate of death from actually having the operation. This means that what we quote as the risk of surgery is an underestimate: by accepting a place on the waiting list instead of having an operation as soon as possible, the patients are taking a risk at least equal to that of the operation merely by waiting for it! Worse still, not all patients waited a whole year. I calculated that if 100 patients all waited a whole year, up to six could die while waiting. On top of that, many of those who did not die waiting did not emerge unscathed. For every patient who died on the waiting list, three others had a heart attack while waiting or became unstable and were admitted to hospital as an emergency, which meant they ended up with a riskier operation and, if they survived, a less good long-term result.

So why do we have waiting lists? Why do we not go ahead and book the patient for an operation at a time that suits him or her when the decision has been made that it is the right thing to do? The answer is simple: insufficient resources. We simply do not have enough ward beds, intensive care beds, operating theatres, nurses or surgeons to satisfy the demand. This creates the wait for an actual operation, but as the above story about Frank clearly shows, it is not just the actual surgery that has built-in delays. Delays are everywhere in the NHS, from getting to see a GP to convincing the GP to refer you to a specialist, to seeing the specialist, to having the necessary tests booked, to having the tests carried out, to the referral for a surgical opinion and finally to seeing the surgeon in clinic: every step of the way is punctuated by delays.

What we need is proper funding for a health service fit for the third millennium, which today means at least 10 to 12 per cent of gross domestic product, or about a 50 per cent increase in the current level of NHS funding. That is what we need. And what do we get? Repeated top-down reorganisation of the service with every new Minister of Health and never-ending demands for more 'efficiency'.

When I first investigated the impact of unduly long waiting times on patients, I wondered if the day would come when patients or their bereaved relatives would start to sue the hospital if it failed to provide timely treatment. After all, it would not be difficult for a medical litigation lawyer to prove that the delay was directly responsible for the poor outcome. It simply did not happen. Perhaps that is because members of the great British public tend to be a stoic lot and, by and large, understand that the health service has many other patients to worry about, over and above their own personal case, no matter how difficult their own health situation. Perhaps they find the idea of suing the hospital that is, after all, trying to treat them and make them better, a distasteful proposition. Perhaps the thought never even crossed their minds. Whatever the reason, I have never received one complaint, let alone a lawsuit, from a patient who was damaged by waiting too long for treatment. Our failures tend to be accepted with remarkable equanimity and grace by patients and their family members, like Frank's wife.

But that looks as if it is about to change. I have a small medicolegal practice, in which I give an expert opinion on whether there was negligence in the management of a heart surgery patient. The work provides some useful extra income, but also gives me an excellent insight into where things can go wrong in my profession; and, just like the disasters presented in McKlusky's Club, it provides me with a valuable opportunity to learn from the mistakes made by others. In the last few months alone, I was asked to adjudicate on three instances where patients died *before* receiving cardiac surgery, and the allegation by the claimant in all instances was that the deaths were due to negligent and undue delays. Of course, I do not know whether these cases will actually reach court and, if so, whether they will be successful, but there is no doubt that the tide is turning. Patients are no longer willing to put up with interminable delays for treatment and, if it takes the threat of legal action to spur the politicians to fund the NHS properly, so be it.

CHAPTER 13

Will *you* do the operation, Doctor?

Heart surgery is not purely about what happens in the operating theatre and the intensive care unit. In elective operations nothing would happen without, first, the cardiologist having referred the patient to a surgeon, and, second, the surgeon having seen the patient in a clinic to ascertain whether an operation is a good idea and whether the patient actually wants it.

Having decided that yes, it might be in the interests of a particular patient to be cut open and have his or her heart fixed, the next step involves explaining the benefits and risks. The benefits are relatively easy: relief of angina; relief of breathlessness; protection from future heart attacks and an early death; and so on. The risks are a little bit harder and need some thought. Most surgeons use EuroSCORE to predict the risk of death — a quick online calculation in which a few questions are answered about the patient, the heart and the planned operation, and the calculator instantly produces a percentage figure for the risk of death. Feel free to try it if you are curious about your own risk of heart surgery at www.euroscore.org. Most surgeons quote the risk directly from the EuroSCORE calculator, but wiser ones make an adjustment for their own figures, tweaking it up or down, depending on how they themselves have been performing in recent years. We then quote the likely risk of stroke, as it is yet another important factor in the patient's decision, as well as common risks associated with any operation, such as bleeding, infection and so forth. Finally, we quote any risks which may be specifically attached to the particular planned operation. Beyond that, it is truly difficult to know what else to disclose. We could, of course, go so far as quoting every possible risk under the sun, including those which are so infinitesimally small as to be ridiculously improbable (such as being run over by a delivery truck in the hospital car park, falling out of hospital windows, and being set on fire by the surgeon), but my feeling is that, if we go this far, the patient will run for the hills. By and large, cardiac surgeons are very honest with their patients and some, like me, brutally so. I do not mince my words when I mention that death and stroke are a real risk, using words such as 'You have to realise that this operation might kill you: there is a 1 in 100 chance that you will leave the hospital dead or seriously damaged.' Of course, in many of my patients, *not* having the operation carries an even bigger risk, so I tell the patients about that too.

At the end of the consultation I usually ask the patient and the family if they have any questions left unanswered. The most commonly asked question is 'How long will I have to wait?' and, in the dire state of the NHS now, that is a particularly difficult one to answer. As I said in the previous chapter, in the bad old days of the early 1990s waiting times of well over a year were quite common, and the death rate from waiting was substantial. Then the Blair-Brown government took over and, for the first time in its history, the NHS was — almost — properly funded. Waiting lists and waiting times shrank right across the country and in some places and specialties, disappeared altogether. Now, with austerity and penny-pinching in all public services, the bad old days are coming back. Some of my patients again have to wait up to a year and, once more, we are beginning to see patients suffer from these excessively long waits. I am now unable to answer this question in any satisfactory manner, and have to resort to an utter cop-out: 'We aim to treat all patients within about three months, but I'll be honest with you and say that we don't always achieve that.'

The second most commonly asked question is 'Will *you* do the operation, Doctor?' and that, too, is a difficult one to answer. What actually happens is that the senior surgeon hardly ever does the operation in its entirety. In most heart operations, opening the chest, joining the patient to the heart-lung machine and closing the chest at the end of the operation are done by experienced trainees who have done this hundreds of times, and they do all of that on their own.

The senior surgeon is usually present for the essence of the operation, the bit in which the heart itself is fixed. That presence could be either as primary operating surgeon, doing the cutting and stitching, or as a supervisor and first assistant to the trainee, who actually does the cutting and stitching. In my practice, about half of the operations done in my name are actually done by a trainee with my direct help and supervision. Does that matter?

Fortunately, it does not. The old perception of the gifted and skilled surgeon who has abilities others do not have may be an attractive and glamorous one, but it is no longer true. Firstly, heart operations are never done by one individual. In a typical coronary artery bypass graft operation (or CABG) there is a team of around 10 people in the operating theatre: a senior anaesthetist and a trainee put the patient to sleep, insert the breathing tubes and monitoring lines; and they administer any drugs needed during the procedure. They are supported in this by an Operating Department Practitioner. The heart-lung machine is run by two perfusionists.

At the leg end of the patient, a Surgical Care Practitioner (or SCP) takes a vein from the leg to use as a coronary bypass. SCPs are not even medically qualified as doctors: they hail from all parts of health care, but are specifically trained in the technical and surgical skills necessary to perform this task. When I started my own training in heart surgery, this was the job that was delegated to the most junior surgeons on the team and we often botched it until we had acquired some experience. Nowadays it is done by experienced and skilled SCPs with a portfolio of hundreds of cases behind them; and, if a junior surgical trainee needs to acquire experience in this, it is the SCP who provides that training.

At the chest end, there is usually a consultant surgeon and a registrar, who is a senior trainee, doing the bits on the heart. In a CABG operation the registrar, within a year or two of starting specialist training, will have acquired the ability to open the chest, take down the mammary artery and insert the pipes in preparation for the heart-lung machine, and will be able to do so unaided. No matter how experienced and confident the registrars are, they are all taught that the first 'surgical instrument' they must master is the telephone: if anything unexpected is found or actually happens when the senior surgeon is not physically present in the operating theatre, they must use the telephone to call for help and support, and we are never more than a minute or two away.

When the preparations are complete and the time has come for the nitty-gritty part of the procedure, how much of the actual cutting and stitching of the bypasses themselves is then done by the registrar depends on three factors: the registrar's level of ability, the consultant's willingness to train, and the level of technical difficulty posed by the coronary arteries themselves. In almost all of my CABG operations, the trainee will do at least some cutting and stitching, and, if conditions permit, the trainee will do most of the technical work. The one thing that I never delegate to a trainee is the decision-making: what to bypass, where to bypass and how to do it. There, I tend to cling to total control over what happens in my operating theatre until I feel that the trainee is now not only as good as (or often technically better than) I am, but also that he or she will make decisions that are as good as mine. Only then, towards the very end of their training years, do I allow them to 'fly solo'. By now they will be ready for a consultant post and actively seeking such an appointment to move to and start their career 'properly'. Until they reach this point, they may well be doing some or even all of the cutting and stitching, although, in reality, I am still doing the operation, but sometimes using their hands.

There is no choice but to train. If we do not, the specialty will die out, and, from a purely selfish viewpoint, when I retire and need a CABG, there may be nobody around to do it. It is, however, important that the training of junior surgeons should not damage patients. To that end, we audited all CABG operations in which a junior was the primary operator as in the above model, with the senior assisting, and compared them with those in which a senior surgeon was the primary operator. It was a major study of over 2,700 patients. We found absolutely no difference in outcomes. The system that we have at Papworth therefore works and is safe.

So: will *you* do the operation, Doctor?

My stock answer is the following: 'My team and I will do the operation. It takes many people to do it, but I promise you that I will be part of the team and I will be there.'

CHAPTER 14

Brazilian

There are precious few advantages to getting old in any walk of life, and that also applies to surgery. Age brings with it bad backs and sore necks, eyesight that is less sharp and a rising intolerance of standing for hours on end at an operating table. One advantage of the passing decades, however, is the multiple opportunities, over several years, of becoming an increasingly 'senior' surgeon, to arrange one's professional timetable in a way that is a little more conducive to leading something of a normal social life. An essential part of such normality is to have at least some down time at weekends. To achieve that, one has to reduce as much as possible the likelihood of being disturbed by the sort of clinical problems which necessitate a return to the hospital or, even worse, a return to the operating theatre. Over the years I have gradually managed to 'front-load' my working week with a lot of operations done on Mondays and Tuesdays, so that most of the inevitable problems that follow major heart surgery have already happened, been dealt with and sorted by the time the weekend arrives.

Despite that, we, as a group of surgeons, all try to use every available operating theatre space on every single working day, and even also at weekends. This is partly to give timely treatment to our patients and to reduce the time that they spend on waiting lists, but also to maximise the efficiency of our service by not having operating theatres stand idle. This means that when some of us are away on leave, the others who are here tend to take over and use their empty operating slots whenever possible. I am always more than willing to take Wednesday morning slots and will often accept Thursday slots, but my enthusiasm wanes when it comes to operating on Fridays. This is for the purely selfish reason that a Friday operation is more likely to disrupt my weekend.

When the surgical patient booking officer called to offer me a Friday afternoon operating slot in a week's time, I was not at all keen. The booking officer explained that she had tried all the other surgeons and none would take this particular slot for a variety of reasons, at least one of which may have been related to its timing at the very end of the working week. I hated operating on Friday afternoon, but I hated the idea of an operating slot lying fallow even more. I had already arranged to go out to dinner with my partner Fran that night, so I reluctantly agreed to operate, but asked her to find a straightforward patient from my waiting list who was having a low-risk, elective, two-to-three-hour operation to be brought in. The booking officer identified a healthy young man on my waiting list who just needed his aortic valve replaced. I readily agreed that he would be eminently suitable.

Unfortunately, the booking office omitted a rather important detail: the young man himself was not informed. When he did not turn up as expected at eight o'clock on Friday morning, somebody telephoned him and it was quickly apparent that he had no idea whatsoever that we had scheduled him for surgery. He had already made other plans for the weekend, which were not at all compatible with having a heart operation at a moment's notice. He politely declined our offer of surgery that afternoon. The booking office called me on Friday morning to explain the fiasco. My immediate response was that I had done my best to 'fill the slot' and that, for once, we could perhaps live with an unused theatre slot at Papworth. But the booking officer informed me that there were several patients in scattered beds around the hospital awaiting urgent surgery, that there was no space for them in the operating theatre for the foreseeable future, and that it would be so, so helpful if I at least operated on one of them that day.

'Fine,' I replied, 'but only the most straightforward case.'

She called again an hour later. She had indeed identified the most straightforward case, but it was not the most urgent. That distinction belonged to an 80-year-old man who had come in with a heart attack, had a weak heart and very severe coronary disease. What's more, he was having dreadful angina just lying in his bed, and that means he was heading for another heart attack. Would I please at least just take a look at him? Please?

The pressure was mounting. Feeling a little less than enthusiastic, I went to see the patient. He was as described, having recently had a heart attack and given an angiogram which showed critical and dangerous coronary disease. He would need all of five coronary artery bypass grafts to sort him out, and if this was not done soon there was a real risk he would not last the weekend. What's more, he was a really nice man with a charming and supportive family round his bed. I know that these factors should not influence clinical decisions, but they undoubtedly do. I agreed to go ahead that Friday afternoon. I informed the registrar who was allocated to me that there was no chance of her actually doing the case: the patient was old and sick and needed too many grafts for a trainee to do it, I declared. I did not declare that I also had a vested interest in finishing the operation quickly, so as to make it in time for dinner.

After some delay, which is not unusual when operating lists are changed at the last minute, the patient was brought to the anaesthetic room and put to sleep by one of the senior anaesthetists. He was wheeled in to the operating theatre and, unusually for me, I was already scrubbed and I started the operation myself right at the beginning. More often, I allow my registrar to open the chest and prepare the connections to the heart-lung machine and I only saunter into the operating room when all that has been done and the vital part of the operation is about to begin, but this time I was in a hurry. I cut the skin with a scalpel and cauterised all the tissues to the breastbone. I then sawed through the bone with a power saw from bottom to top and the chest was opened. I placed a retractor which raises the left half of the breastbone to expose the mammary artery which lies on its inner surface. With a forceps and cautery, I dissected this artery out until it fell off its attachment to the chest wall and lay free, ready to be joined to the left anterior descending coronary artery (or LAD), which is the most important bypass target.

There is something strange about the left internal mammary artery (or LIMA). It leaves the subclavian artery just behind the collarbone and travels inside the chest and just in front of the heart alongside the left edge of the breastbone, all the way down to the diaphragm between chest and belly, where it divides into small and somewhat insignificant branches. It was first used as a bypass graft in 1964 by a pioneering and visionary Russian heart surgeon called V. I. Kolesov, who joined it to the all-important LAD on the front of the heart. After many years of being ignored by surgeons in the West, the LIMA has finally gained its rightful pride of the place as the best tube there is to bypass the LAD, and for very good reasons, too. It's these reasons that are a bit strange.

Whenever we do a CABG, if the LAD needs a bypass we almost always choose the LIMA as the bypass tube. This is because of the following:

- It is in the vicinity. When we dissect it off the chest wall, it will easily reach any part of the LAD that we want it to reach.
- It seems to be superfluous: when we take the LIMA away from the chest wall, nothing happens. The chest doesn't fall apart and the patient suffers almost no consequence from its removal.
- It happens to be about the same size as the LAD in most people, so that joining the two together is easy, and it will provide the right amount of blood flow. If by some chance it is smaller than the LAD, it will grow to match the size of the LAD within a week.
- It seems to have magical protection from atheroma, the disease that causes arteries to fur up with cholesterol. No matter how old the patient is, and no matter how awful the atheroma is elsewhere, and no matter how many parts of the body are affected, the LIMA is always whistle-clean. I have seen patients with severe atheroma furring up all of their heart arteries as well as the arteries to their brain, the ones in their belly and the ones in their legs. Despite this, their LIMAs are always healthy and free of atheroma, and can be used as a bypass.
- Finally, a bypass made by joining the LIMA and the LAD also appears to have magical protection: provided it is done competently, the bypass graft will flow for the rest of the patient's life. In fact, the patients can be almost 100 per cent sure that they will meet their maker with that bypass still functioning.

When you read all of the above, it is tempting — especially if you are a creationist — to take the existence of the LIMA as proof of the existence of God: you could easily argue that the LIMA was intelligently designed to serve as a bypass graft to the LAD, since it fulfils that function so incredibly well. Perhaps it was put there just so that my surgical colleagues and I can make use of it in coronary bypass. (If you are an atheist, don't worry, as you could also argue that a far easier and more intelligent piece of design to solve the problem of atheroma in the coronaries would be for us not to get atheroma in the first place. You would, of course, be right.)

The rest of the bypass grafts would be built using veins from the leg, a good option, but nowhere nearly as good as the LIMA to the LAD. Essack, one of our surgical care practitioners, was harvesting these veins at the same time as I was preparing the internal mammary artery. I then opened the pericardium and looked at the heart. It was covered in bright yellow fat and was beating somewhat sluggishly and not too elegantly, but it was an 80-year-old heart, after all.

I placed the pipes in the aorta and right atrium to connect the patient to the heart-lung machine, and asked Essack if the vein from the leg was ready for use. He said it was, and I said, loudly and clearly, 'On bypass, please.'

The perfusionist started the heart-lung machine. Blood flowed from the right atrium into the machine and came back, oxygen-rich, into the aorta. The anaesthetist switched off the lungs. I had a quick look around the heart and identified where in the coronary artery network the five bypass grafts were going to be attached, and then applied the clamp to the aorta, isolating the heart from the circulation. I gave the protecting potassium solution to cool the heart and make it stand still while it received no blood supply. I took the piece of vein from the leg and joined it to the first coronary artery to be bypassed, using a very fine polypropylene suture. I gave some more potassium solution down the graft to ensure that it had good flow, and that there were no leaking points at the join. I then trimmed the vein to the right length and did the second join, then the third and fourth. Four coronary arteries now had four ends of pieces of vein joined to them beyond the blockages, but the other ends were still lying free.

Finally, I joined the end of the mammary artery to the LAD. The part of the operation that required isolating the heart from the circulation was now over, so I removed the clamp. Blood flowed down the patient's own coronary arteries, washed away the potassium solution and, slowly, the heart began to beat again.

All that was left to do now was to connect the four pieces of vein to the aorta, and the bypass operation would be complete. For that, we apply a partial clamp on to a bit of aorta to allow us to work without blocking the aorta completely. I applied the clamp, did the necessary stitching and the four vein grafts were now functional. The entire procedure on the heart itself had taken 78 minutes and it had been a good and slick operation. The anaesthetist switched the lungs back on and 'Off bypass, please,' I asked. The perfusionist stopped the heart-lung machine and the heart took over its old job with no difficulty whatsoever. Just a bit of 'tidying up', placing drains and pacing wires and we would be ready to close. I felt virtuous because I had given the patient a timely and possibly life-saving operation and selfishly pleased for having done this so quickly that I would be able to make my dinner date.

Life was good.

Just as I was about to begin closing the chest, there suddenly appeared to be a lot of bright red blood. It was coming from one of the joins between leg vein and aorta. 'I must have botched that join,' I thought, and irritably asked for a suture to fix it. I was just placing the suture when a second join next to it began to bleed, even more profusely. I stopped. I could not remember having any difficulty with this part of the operation, so why were two of my four joins bleeding so much? While I was busy pondering this, the third and the fourth also started to bleed. Horribly. With all four joins hosing, it was a blood bath. Then illumination dawned. I asked the anaesthetist to insert an echo probe down the patient's gullet to look at the aorta, and it confirmed my worst fear: there was acute aortic dissection. The layers of the aorta were separating before our very eyes as we were looking at the echo images and the aorta itself. This was a real heart-sinking moment.

Denial. Isolation. Anger. Bargaining. Depression. Acceptance.

There was only one thing for it: back on the heart-lung machine, replace the ascending aorta, and, to add insult to injury, redo the entire four vein joins, since the bit of aorta to which they were attached was now fit only to be discarded. Unfortunately, the heart-lung machine can no longer be connected to the aorta, as the latter is now in shreds, so another artery must be found (and found quickly, since the bleeding was continuing apace and barely controlled by the registrar's two hands compressing the aorta with large swabs). The most accessible artery for such desperate situations is the femoral artery in the groin. I cut a circle out of the drape overlying the right groin and asked for some surgical prep solution to sterilise the skin. I splashed it on the area, took a scalpel, cut across the groin and proceeded to cauterise some minor skin blood vessels. The patient's groin went up in flames.

Total panic ensued. The flames were blue and orange and were already beginning to set fire to the surgical drapes in the few seconds it took me to realise what had happened. The prep I had used was 70 per cent alcohol, not one that we would normally go for when cautery is being actively used, but I had stupidly neglected to check the solution and, with massive haemorrhage at the chest end, I was in too much of a rush to wait for the alcohol to dry. The flames were almost reaching the theatre operating light before we finally managed to extinguish the blaze by smothering it using whatever was available in the vicinity, including sterile swabs and towels and the surrounding surgical drape material.

With the incendiary crisis over, I carried on. I found

the femoral artery, joined a tube to it and we went back on the heart-lung machine. The tattered aorta was clamped again, more potassium was given to stop the heart, and I cut out the ascending aorta. I used surgical glue and Teflon to reconstruct the layers of the aorta at the cut ends, and then sutured a Dacron graft to replace the part I had excised. I finally redid the four vein joins on to the Dacron graft.

The original operation had taken 78 minutes on the heart-lung machine, during which the heart was isolated for a mere 45 minutes. Repairing the aortic dissection had taken 200 minutes on the heart-lung machine, during which the heart was isolated for a whole 80 minutes — and the poor old heart, after all this abuse, understandably somewhat struggled to take over the circulation this time. It finally managed to support the circulation with the help of some drugs and 'tincture of time', but the restaurant to which we were going for dinner had long closed by the time I was able to leave the hospital.

The following day, the patient was making excellent progress, and the entire surgical team and I were delighted with this amazing 'save'. The hospital management, however, was less than delighted, and did not see things the same way we did. As far as the managers were concerned, the singular and most salient feature of the entire episode was that one of their senior, experienced surgeons had set a patient on fire, and that would simply not do at all.

In March of 2015 the NHS published a list of 'Never Events'. These are things that the NHS considers should never happen in health care, and quite rightly so. They include horrendous and unthinkable incidents, such as 'Never Event Number 1: Operating on the Wrong Site' — for example, taking out the healthy kidney and leaving the diseased one, or doing the right operation but on the wrong patient. These highly unlikely catastrophes do happen in health care, albeit very rarely, and when they do, thorough investigations are conducted and heads may roll. Interestingly, the 'Never Events' list also includes 'Never Event Number 10: Falls from Poorly Restricted Windows' (see Chapter 4). The relevant item here, however, is 'Never Event Number 14'. It is 'Scalding of Patients' and is defined as the patient being scalded by water used for washing or bathing, but this specifically excludes scalds from water being used for purposes other than washing or bathing (such as from kettles or teapots). I tried very hard to convince our clinical governance team that my setting fire to the patient did not really qualify as a 'Never Event' and was not worthy of a massive investigation, because (a) no water was involved, and (b) nobody was bathing the patient anyway. But this did not go down well with my managerial colleagues and a full investigation was nevertheless carried out, with several recommendations aimed at preventing such an event ever happening in the future. This was, of course, the right thing to do.

In the meantime, the patient himself made a good recovery from both his surgical and incendiary tribulations, and he was bemused, a few days later, to find that he had not a single pubic hair left as a result of the accidental 'Brazilian' by flames. We candidly explained the events to him in detail and I pointed out that he could sue us if he wanted to, but he was happy with the end result and took no further action.

CHAPTER 15

Keyhole surgery and other novelties

Elsewhere in this book I have related the stories of two patients who had a minimally invasive coronary bypass operation. Both of them had suffered terrible complications as a result. The problems that arose were, in neither patient, directly related to the minimally invasive approach and could have happened in any CABG. In fact, when executed carefully in the right kind of patient and by the right kind of surgeon, minimally invasive surgery can work remarkably well and the patient is usually delighted both with the result and with the fast recovery. That is not always the case.

The first time I ventured into minimally invasive CABG was a bit of a gamble. I took a big risk and, more importantly, so did the patient. I still remember it vividly as it was Christmas in 1996 and I was on call for the entire holiday period. Papworth Hospital was festooned with seasonal decorations; Christmas trees seemed to be everywhere and many nurses and doctors wore festive baubles and bits of tinsel as they went about their work. As always at that time of year, the hospital was gradually filling up with patients needing heart surgery with varying degrees of urgency: peace on earth, goodwill to all, and open heart operations for the few who cannot be discharged to await elective surgery. Many patients were old and some were very sick. While the rest of the nation was busy finalising the preparations for the indulgent feast of food, drink and dire goggle-box offerings that seem to constitute modern Christmas, the cardiac surgery teams at Papworth had other preoccupations.

One of our more difficult tasks was to prioritise these patients, so that the most critical ones received their operations first. The rest would have to wait for whenever there was space in the operating theatres and an availability of beds in intensive care. Amongst the patients in question was a relatively young man with angina so troublesome that he could not be sent home with it, and his tests had shown a single total blockage of the LAD, which is the coronary artery at the front of the heart. This was not a life-threatening condition, but we still had to fix him before he could go home.

It happened that a few weeks earlier I had attended an international conference on heart surgery. As is often the case, many of the research and clinical papers being presented were predictable and unexciting, but one talk in particular caught my attention: Dr Valavanur Subramanian, a New York heart surgeon, was describing his new method of performing a single CABG to the LAD: he would make a small cut on the front of the chest, go between the ribs, and then he would bypass the blockage by joining the left internal mammary artery to the LAD, while the heart is still beating: no splitting the chest, no heart-lung machine and no stopping the heart. He had done this in quite a few patients with good reported results. It seemed to me that this was a simple and most attractive proposition and I had thought 'I'd really like to try that one day.' When I saw that young man on Christmas Eve, it struck me that he would be an ideal candidate for this minimally invasive or so-called 'keyhole' approach.

I spoke to him in general terms about the risks and benefits of having a CABG and he was very keen to go ahead, as he was exasperated by the angina and wanted to get his quality of life back. Then I talked to him about the keyhole approach as an option, despite the fact that my experience of the procedure was non-existent. It is true that in those days, much surgical training could be summarised as: see one, do one, teach one. Can you possibly countenance an airline pilot being trained in this way?

I was totally honest with the patient: I told him that I had never even seen this procedure done before, let alone done it myself, but added that I had listened to a New York surgeon talk about how it should be done and that it did not look all that difficult. I promised him that I would do my best not to subject him to any additional risk and that I would immediately switch to a conventional open-chest operation if we encountered the slightest problem whatsoever. I also suggested that if he was at all apprehensive or uncertain, we would, right away, forget all about the minimally invasive approach and go straight to the conventional operation. His response surprised me: he rather too readily agreed to be a 'guinea pig', and in fact appeared more enthusiastic about the minimally invasive prospect than I was.

The operation was booked for Boxing Day. I explained to the operating team that we were going to try something new, and then scrubbed and donned gloves with a fair bit of trepidation. I carried out the incision and separated the mammary artery from the chest wall, ready to join it to the LAD. Then I opened the pericardium and the LAD was right there: visible and accessible. The join took no more than 12 minutes, despite the fact that the heart was beating. It was so easy! Nowadays we do these operations with a lot of help from technology, such as stabilisers and suction devices to keep still the bit of heart we are working on, and carbon dioxide blowers to allow us to see what we are doing by dispersing any blood that blocks the line of sight. I had nothing of the sort, as these devices had not yet been invented, and somehow managed to do the operation with a basic surgical set, not far removed from a knife and fork. We closed the wound and the patient made a rapid recovery, going home three days later with his angina gone. He was thrilled, and of course so was I. In my mind the decision was firmly made to continue with this procedure and do it many more times, readily offering it to any interested and suitable patient like him over the subsequent 20 years.

To be honest, I was really lucky on that Boxing Day. I have since done some 200 such operations over the last two decades with mostly excellent results (the two awful ones are described in detail in this book), but I can absolutely testify that not a single one of them had been anything like as easy as the first. Some have been really fiddly and difficult, and, had my first attempt been one of those, I may well have given up on the procedure there and then. In fact, that first operation in 1996 was the first of its kind to be done in the UK, and I had precipitously plunged into it with minimal or no preparation, helped by the prevailing culture at a time when doctors and surgeons had a huge amount of clinical freedom to do whatever they wanted with relatively little oversight and with nobody daring to question them. If I were to do that again now, I would probably be subjected to a serious hospital investigation and, if things went badly for the patient, I might even lose my job, perhaps even lose my registration with the General Medical Council and never practise medicine again.

The subject of innovation in medical care is a thorny one. Every time something new is tried, there are potential risks attached, and, if the treatment is a truly novel one, these risks may be totally unknown. Nowadays, before a new drug is released for unrestricted use on patients, it must be put through an almost unbelievable panoply of tests under the strictest observation possible, from biochemical and pharmacological investigations to ridiculously detailed testing on animals, before a limited and very rigorous pilot trial begins on actual patients, looking with an eagle eye focused on every conceivable and inconceivable side effect. The process takes many years and a drug may fall at the first hurdle or at any point thereafter. If it does, the millions which will have been invested into its development and testing will have gone to waste, and the researchers go back to the drawing board and start again. Most novel pharmaceuticals are thus tested to destruction before being introduced into clinical use, and the days of catastrophic scandals, like the infamous thalidomide affair, are well behind us.

Not so in surgery. The problem with new surgical techniques is that it is very difficult to have a system which follows and monitors every new manoeuvre that a surgeon decides to try on the mere basis of a personal belief that it 'seems a good idea'. Surgery is a series of hundreds of steps and a surgeon may decide to modify one, a few or many of the steps at will. This is part of our make-up. Without doubt, our ability and willingness to 'tailor' an operation to suit the patient and the situation at hand can be extremely useful, and sometimes such 'on-the-hoof' modifications of surgery can actually save lives. Most of the time we deliver well-tried-and-tested procedures in a tried-and-tested manner. Sometimes we have to be inventive and, rarely, we may innovate in a radical fashion. Such behaviour can help patients, but sometimes it can do them a lot of harm and even bring their lives to an abrupt and premature end. Fortunately, most surgeons are sensible and, if they do innovate, they also do that sensibly, but not all.

There is no doubt that many of the huge advances in heart surgery were built on the dubious foundation of the almost wanton risking of patients' lives. Fifty years ago this was undesirable, but perhaps forgivable. After all, desperate patients whose quality of life was appallingly limited by their symptoms, and whose life expectancy was appallingly limited by their heart disease, did not have a lot of choice. It was very much a question of 'do or die'. When heart surgery started, its early targets were just such types of patient. With that in mind, we can understand why surgeons would 'have a go', devise a new procedure to test on the hapless patient and take a phenomenally large risk in doing so: the patient often had no other viable choice.

Thus, when, one day, John Bailey sped across Philadelphia to fix his second mitral valve of the day, scheduled in a different hospital precisely so that the authorities would not stop him from doing it (having heard that his morning patient had died elsewhere), he was displaying the spirit of impetuous, trail-blazing inventiveness and supreme confidence that characterised many of the pioneer heart surgeons of his generation. This spirit was also very much alive when Vasilii Kolesov in Russia carried out the first coronary artery bypass graft, when Christiaan Barnard in South Africa carried out the first heart transplant, and when Nina Starr Braunwald in the United States first replaced a mitral valve. Of course, for every brilliant triumph of this type, there were, in the shadows, countless dark disasters and catastrophic failures, which led to patients being maimed or killed, but most of these patients were already on death row, because of their heart disease.

Nowadays, heart surgery and its routine techniques are so well established, and so safe, that a new procedure must hold a substantial promise of significant improvement over and above what is already available in order to justify subjecting patients to the risk associated with the new and the unknown. For many innovative surgeons, especially those who carry within them at least part of the spirit of their predecessors, this is a problem, and that makes it a problem for the patients.

The famous Italian thoracic surgeon Paolo Macchiarini was recently the subject of an intensive investigation into his novel method of replacing tracheas (windpipes). The trachea is a very important organ: it is the vital pipe through which air gets into the lungs. If it becomes narrowed, breathing turns into a laborious nightmare. If it becomes blocked, death by suffocation is the result. That is why the Heimlich manoeuvre saves lives by dislodging food that blocks the trachea when it goes down 'the wrong way'. Fortunately, diseases and tumours that cause tracheal narrowing are few and far between, but they do exist. The commonest cause of a tracheal problem is the scarring that results from having a plastic tube in the trachea for a prolonged period of time, such as when someone spends weeks on the intensive care unit after a major accident or a serious illness.

We can replace so many parts of the body with artificial substitutes: from heart valves and arteries to hip and knee joints. All of these are internal body parts living in a clean, bacteria-free environment, a relatively safe place to implant a manufactured replacement. The reason we cannot do this to the trachea is simple: it is a tube where the internal body tissues surrounding it on the outside are clean and sterile, but the inside is dirty: it is constantly exposed to air and bacteria in our surroundings, inhaled with every breath we take. This means that replacing part of the trachea with, say, a plastic tube would immediately expose the surgical joins to bacteria. Our body's immune system is pretty efficient at killing bacteria that find their way into healthy, live tissue, but if bacterial infection settles on an area where there is foreign material right alongside the live tissue, you can expect big trouble. No matter how hard the immune system tries, its cells and antibodies cannot reach the nooks and crannies of a piece of plastic to sterilise it. In fact, the standard correct treatment for any piece of infected prosthetic material implanted in the human body is to remove it as soon as possible so that the immune system can stand a chance against the infection.

Macchiarini decided to replace tracheas with plastic. He thought that by bathing the plastic in bone marrow extract full of stem cells, a coating of live tissue would develop and that this coating would shield the artificial trachea from infection by making it more akin to live tissue. He probably did not test this theory with sufficient rigour and he was subsequently accused of falsifying his published papers on this pioneering procedure, or at least of being somewhat economical with the truth when describing the complications that ensued. Nevertheless, he had published papers 'proving' the success of his method in prestigious journals and had arrived at his new post in the Ear, Nose and Throat Department at the world-famous Karolinska Institute in Stockholm with a reputation as a world-class pioneer and something of a miracle-worker.

He carried out some operations with what looked like promising results in the early days. When he needed to use the heart-lung machine to do an extensive tracheal replacement, he asked for access to the facilities in the cardiac surgical service. He operated on two patients there and both patients died. The director of cardiac surgery, a close friend of mine called Ulf Lockowandt, immediately stopped him from doing any more such cases in his department.

Gradually, and after several investigations, the truth was out that most of the patients on whom Macchiarini had tried this procedure in many countries around the globe had died horribly, and that none of the very few survivors had any quality of life to speak of. The story of this surgeon and his unfortunate patients has since been chronicled in a BBC documentary and it makes harrowing viewing.

Macchiarini practised thoracic surgery, the sister specialty of heart surgery, but some of his excesses can also be seen in my own specialty, and probably in many other areas of surgery. One of the innovations that are captivating surgeons (and the public) in every specialty is a desire to perform operations with minimal incisions, so-called keyhole or minimally invasive surgery. Heart surgery is no exception, and in recent years there have been many developments aimed at completing heart operations through tiny cuts. It can, of course, be done, and many heart surgeons round the world have done it brilliantly. The problem is that, as a surgical technique, minimally invasive heart operations have become 'sexy', and with that, caution has been thrown to the wind.

One of the operations that has been modified to be minimally invasive is mitral valve repair. In the traditional operation, we split the breastbone from top to bottom, put the patient on the heart-lung machine, clamp the aorta to isolate the heart from the circulation, and then open the left atrium to look directly at the mitral valve. We assess it, test it to see what is wrong with it and repair it using a variety of methods. The operation is hugely successful and very durable, and the rate of death from the procedure is among the lowest of any heart operation: well below 1 per cent. Mitral valve repair is already a success story, hard to improve upon and therefore a tough act to follow.

When a mitral repair is done through a minimally invasive route, a special tube is placed in the windpipe to allow the right lung to collapse so that the heart can be seen through the telescopes inserted into the right side of the chest. The heart-lung machine is still needed, so a cut is made in the groin to allow the machine to be connected to the large blood vessels that are there. A number of small cuts are made in the chest to insert the telescopes and long-handled instruments needed to do the operation, as well as a special extra-long clamp to block the aorta and isolate the heart. Then the operation can begin and the mitral valve can be repaired with these long instruments. The only thing that is truly minimally invasive about this approach is the absence of a long cut on the front of the chest: in almost every other aspect of the operation what is being done to the body is at least as invasive as the traditional operation, and perhaps a little more so. Nevertheless, we know that it can be done this way, so where's the harm? The harm only strikes when you add an additional and highly toxic ingredient, which is surgical ego.

It is readily obvious that a mitral repair done with limited access and with telescopes and extra-long instruments is bound to take more time than one done with full access and under direct vision. A sensible surgeon, on trying this approach for the first time or relatively early in his or her experience, should have a heightened awareness of the ticking of the clock. The operation on the mitral valve can only begin when the heart is isolated from the circulation by placing a clamp on the aorta. The minute that clamp is placed, the heart begins to die slowly from lack of oxygen. We can, of course, slow this process as far as possible by infusing a cold potassium solution, but even that will not protect the heart forever. The sooner the clamp is released and blood flows into the coronary arteries once again, the better and quicker is the heart's recovery. After four hours of oxygen deprivation, very few hearts will function normally, most will need drugs or machines to help them recover and some may not function at all.

This means that, after a certain length of time has passed, if little progress has been made with the mitral repair and if the procedure looks as if it will take an unacceptably long time, a sensible surgeon should simply decide to change tack. Cut your losses, open the chest and do it the tried-and-tested method, and that way both you and your patient will live to fight another day. Eat humble pie and admit to the patient tomorrow that you had to abandon the minimally invasive approach, and the only casualty will be your pride. Unfortunately, some surgeons do not see things that way. Carried along by a wave of ego, arrogance and grim determination, they persevere, so that it is well known on the grapevine in heart surgery circles that, in many countries as well as in the UK, young, otherwise fit patients have actually died from having this procedure, when their hearts were starved of oxygen for up to 10 hours. Some surgeons have lost their jobs as a result of such mishaps, but others have got away scot-free. The most upsetting feature in this sorry affair is that these patients would almost certainly have survived if their operation had been conducted in the unsexy, traditional but 'boring' old-fashioned way.

These events should serve as a salutary reminder to all of us that the primary motivation of our work should be to help patients feel better and live longer. Success, ego-boosting, fame and fortune may well be highly desirable by-products of our endeavours, but they should never become the primary motivator. And if you are offered a minimally invasive heart operation, after you make sure that the surgeon has done lots of them, with audited results in a hospital with robust governance mechanisms, insist that you value safety above the size of the cut and encourage your surgeon to bail out if it looks like the procedure is not going well and if your heart and your life are being put at risk.

CHAPTER 16

This time it's personal

Monday, as I have said, is my busiest operating day. I may do three consecutive operations or, if not three, two, in which case at least one of the two would be a really big one. Monday operations are rarely finished before eight o'clock in the evening, but that does leave just about enough time for a trip to my local gym after I get home. On the face of it, the last thing you need after a long and arduous day in the operating theatre is a gym session, but surgery, although undoubtedly physical, involves an awful lot of standing in fixed positions, a substantial amount of muscle tension and, sometimes, a fair bit of psychological stress. This means that the surgeon often limps away from the operating table as stiff as a board, and there is nothing like a few gym exercises to loosen stiffened muscles and clear preoccupied minds.

Fellow surgeon Steve Large and I usually do our utmost to make it to the gym on a Monday night, and, during our exertions, we discuss recent cases, research ideas and hospital politics. Having put the world to rights, we normally try to follow that with a visit to the Kingston Arms, a very nice traditional Cambridge pub, for a pint of Jaipur Pale Ale and a burger with all the trimmings, and this, of course, serves admirably to undo all the good that the gym session might have achieved.

To a naturally lazy person like me, aerobic exercise can be mind-numbing and soul-destroying in its tedium, but a *Guardian* cryptic crossword set by Rufus (who was the usual Monday compiler) did wonders in making the obligatory 20 minutes of aerobic workout pass very quickly. Cross-trainer exercise machines have a large screen with all sorts of vital information about your speed, effort, incline, calories burnt, length of time exercising, heart rate and so forth. All of this electronic wizardry is of no interest to me. The screen's best feature is that it is large enough to double admirably as a bookstand on which a newspaper (judiciously folded on the crossword page) can sit most comfortably, and just below that is a shallow ledge on which one can park a pen. That is precisely why a cross-trainer is my favourite aerobic machine.

So there we were, Steve and I, on adjacent cross-trainers, each working out with a copy of that day's *Guardian* crossword covering most of the cardiovascular data screen. As the last minute was mercifully ticking away, I decided, for no particular reason, to intensify the workout and increased my speed, safe in the knowledge that it would all soon be over.

As I speeded up, a rather uncomfortable feeling in my chest began to assert itself. I, of course, ignored it. Unfortunately, it did not ignore me. It got worse. The 20 minutes were up and the machine duly instructed me to 'cool down', so I did, and the discomfort went away after less than a minute. Nevertheless, I was a tad concerned, so I decided to extend this aerobic session and intensify the exercise once more, and, just as I feared, the discomfort returned. Steve had already headed for the weight machines with a throwaway remark that unless I did the same pronto and got on with it we would miss last food orders at the Kingston Arms. I slowed down and the discomfort immediately abated. I speeded up one last time and there it was again. I speeded up further and it became quite unpleasant, so I stopped. This is the classical textbook description of angina.

Fuck fuckety fuck.

I am a master of denial. I said nothing at all of this to Steve. Two pints of Jaipur ale and a cheeseburger later put all of this in perspective. After all, what is one occasion of chest discomfort once in a blue moon? I was feeling fine and I soon forgot all about it until the next gym session when I thought that, just out of idle curiosity, I would try to push myself again and this time, not only did the chest tightness return, it did so at a lower intensity of workout. The following day, it happened when I was cycling up a very gentle incline and I finally decided that the time had come when I should come clean and tell someone. So I said to Fran that evening that I thought I had angina and that it seemed to be rapidly getting worse. We talked it over sensibly and agreed that the only course of action was to seek help as soon as possible, and I resolved to speak to Peter Schofield, a good friend and an excellent and trusted Papworth cardiologist, at the first available opportunity.

The following evening, angina came halfway up a single flight of stairs and I was genuinely alarmed. There is a name for the type of angina that increases sharply in severity in a very short space of time: it is called 'crescendo' angina. At the peak of the crescendo looms a heart attack, unless something is done. The following morning I nabbed Peter in the corridor and he offered to see me formally for a consultation in his afternoon clinic.

Yes, he said afterwards, this is crescendo angina, and yes, it needs urgent investigation. He explained that it could be disease in a single coronary artery, in which case a stent may fix it, or it could be dangerous disease affecting all the coronary arteries, in which case surgery (our old friend CABG) would be required. He suggested that I have a CT scan first, because if that pointed towards a single vessel disease, I could go ahead to a coronary angiogram with a 'view to proceed' to stenting if found suitable. He also suggested that, depending on the CT findings, I may wish to choose my CABG surgeon in advance, just in case.

Bloody hell. This does not happen to me — this is what I do to other people! The role of surgeon-turned-patient appealed like a hole in the head.

Peter started me on a handful of anti-angina medication

and also on a big dose of two 'antiplatelet' drugs. Crescendo angina often means that a plaque in the wall of a coronary artery is unstable and in the process of rupturing. When the platelets in the blood stick to the raw area of a ruptured plaque, a small clot forms and when that clot grows it can block the artery completely and thus precipitate a heart attack. Antiplatelet drugs, like aspirin and a few others, can reduce that risk. I started taking the tablets and did my best to put it all to the back of my mind. After all, there were real patients to see and operations to do, and the powers of denial were at work again.

That evening I received a phone call from Deepa, one of our radiologists. She asked what I was doing on Monday.

'Three cases, Deepa. Why?'

'Oh,' she said, 'Peter wants you to have a CT. Sounds like Monday is a busy day for you. Oh well, not a problem. We'll open the department early just for you — come to the X-ray Department at seven.'

My powers of denial were no longer adequate. This was truly happening.

It was a sunny Monday, if a little cold, when I rode my motorbike to Papworth at six thirty in the morning. As I walked into the deserted waiting room of the X-ray Department, I was confronted by a large poster of a sympathetic-looking doctor in a white coat and bow tie, smiling benignly across his desk at a middle-aged female patient. I did a double-take when I realised the doctor in the poster was me! This was getting surreal. Then I remembered that the hospital had asked me only a few weeks previously to pose for such public information photographs on the pretext that I looked 'trustworthy'. I had dug out an old bow tie, complied with the photo shoot and promptly forgotten all about it. As I was reading the caption of the poster — all about the 'personal and kind care that Papworth offers' — Deepa appeared and instructed me, personally and kindly, to come in and climb on to the CT scanner table.

'You need to be relaxed for this,' she informed me. 'The CT images won't be any good if your pulse rate is 80 per minute or more.'

'Not a problem, Deepa,' I cheerfully replied. 'I'm fit. I go to the gym regularly and my resting pulse rate shouldn't be above 60.'

She hooked me up to ECG monitor, which confirmed, indeed, that my pulse rate was a steady 56 per minute.

'Good,' she said. 'Now the first run will be to look for calcium in the coronaries. If there is none, you are very unlikely to have coronary disease. Once that is done, if there is calcium, we will do the detailed images and focus on the affected areas. The first run should be quick.'

She then stepped out of the room to protect herself from the radiation and spoke to me through the intercom from behind the lead-lined glass screen. The machine whirred into action and the table, with me on it, moved slowly into the tunnel of the scanner. The first run was indeed quick, and as I came out of the tunnel, I could immediately see the images on the monitor in the corner of the room. Like an exuberantly decorated Christmas tree, my coronary arteries had lit up with bright patches of white calcium absolutely everywhere.

Deepa entered the room again, this time brandishing a large syringe.

'What's that for?' I asked.

'It's an intravenous beta blocker to slow your heart down,' she replied. 'Your pulse rate went up to 130 when you saw the images.'

With beta blocker duly administered, the scanning continued, while I for some reason suddenly became aware that I had not made a will since I was in my early thirties, when my first son was born. I clearly had coronary disease and this could be fatal. There was a lot to sort out. A year previously I had, for no particular reason, realised that it was going to be my thirtieth anniversary of working for the NHS and that in the entire 30 years I had not yet taken a single day of sick leave. As the anniversary was approaching, I irrationally began to dread that I would contract some dreadful pestilence or other lurgy just in time to ruin a hitherto perfect health record. The anniversary came and went and I had remained as fit as ever. Had the chickens now come home to roost? My awful family cardiac history and the 20 Gitanes a day that I had smoked and hugely enjoyed until my late thirties had clearly taken their toll. How I'd have loved to light a Gitane that very minute!

The scans were completed. Deepa suggested I go up to her office to review them. Yes, there was a lot of calcium.

Yes, my coronaries were not healthy, but most of the calcium seemed to be in the wall of the arteries and did not appear to be obstructing blood flow, except in one artery, the left anterior descending (or LAD). It is of course the most important single coronary, but at least it's only one — or so the CT scan seemed to suggest.

Peter Schofield agreed with Deepa's findings and proposed a coronary angiogram with 'a view to proceed' to stenting the following Thursday. He wasn't too happy about the calcium, and thought some drilling of the artery may be needed to allow the stent to go in, which is an indication that it would be a difficult procedure and that the success rate would be lower. Because of that, he also again suggested that I find a surgeon, just in case. I had already given the matter some thought and pretty much decided that I would ask David Jenkins to do the necessary if the need arose. David did not do many CABG operations, but he always treated them with respect and was a careful surgeon who hated taking risks — a bit like me, or so I like to think. I went to his office.

'David, I've got a favour to ask,' I said. 'I'm having a stent put in next Thursday and you're around, so would you mind awfully covering the procedure in case CABG is needed?'

David's already pale Welsh complexion blanched to a whiter shade of pale.

'Why, yes, of course,' he replied.

I thanked him and went to the operating theatre to start the first of the day's three cases. Later, I felt a lot more cheerful. I had an undeniable problem, but the steps were in place to find the solution, and, more importantly, it took three flights of stairs to bring on the angina that evening instead of just one. At least things were not getting worse.

I took the following Thursday as sick leave and, in so doing, permanently blotted my perfect health record.

Being admitted as a patient to one's own hospital is a very strange experience. The role reversal is not too difficult to cope with, but it is of course very noticeable. Despite my best attempts at being a docile, model patient and doing what I was told, the nurses, I observed, tended to give me a wide berth, and the young intern was palpably nervous when she took blood for tests and went through the history and consent form. Once the formalities and blood tests were done, there was little to do but wait, and I even wondered if I could nip down the corridor to my office and use the time productively to deal with a few discharge summaries and patient administration tasks.

Very few people in the hospital knew that I was in as a patient, but the few who did popped in to say hello, and my partner Fran came to sit by my bedside and offer moral support. When the time came to be wheeled into the cardiac catheter laboratory for the angiogram I was as calm as one could be under the circumstances.

Peter must have done tens of thousands of angiograms and several thousand stent procedures, including the relatively recent and widely publicised one on Prince Philip, the Duke of Edinburgh. He was slick and efficient. The needle went into my femoral artery in the groin, and a guide wire was pushed through the needle and up towards the heart. Then the angiogram catheter was advanced over the guide wire and manoeuvred to lie in position at the entrance of first the right and then the left coronary artery. The dye was injected, the pictures were taken and I could see them clearly displayed in real time on the various monitors scattered around the room.

Joy of joys: the right and the circumflex coronary arteries looked whistle-clean. In the LAD, there was the culprit: the crater of a ruptured plaque which narrowed the artery, but by no more than 20 per cent — hardly at all. Clearly the problem was already in the process of resolving itself, the antiplatelet drugs had dissolved the clot or at least prevented its growth, the rupture would heal over time and I would be left with very little, if any, narrowing.

'I think we can afford to leave this alone,' said Peter, with a smile on his face as he took off his mask and gloves.

I agreed enthusiastically with a sense of huge relief. To make full use of my first day of sick leave, a mere two hours later Fran and I were in the garden of a gastropub in a nearby village, celebrating.

The angina continued slowly to get better after that, but it took nearly two more years before it disappeared completely. I still always watch out for it when cycling uphill or while working out at the gym. I now also know for a fact that, like so many men and women of my generation, I have coronary disease, and that one day the angina may return as the disease progresses. I have much more empathy with my patients who have angina and especially with their heightened level of worry and concern when they are told the diagnosis. I also know that if I ever did need a stent or CABG I would not be deliriously happy with the prospect, but I would go ahead willingly. If I live long enough, it will only be a matter of time.

Insider knowledge can sometimes be terrifying. Paradoxically, it can also be very reassuring.

CHAPTER 17

The many forms of Lazarus

I have just this minute returned to my office from the cardiology ward.

A mere 25 minutes ago I had made myself a cup of coffee and was just about to sit down at my desk and assess some patient referral letters when I received a phone call from the ward. The call was from a doctor in the cardiology service. He told me that a patient's heart had stopped, that the cardiac arrest team were there and that major surgical interventions were likely to be needed, so could I — the surgeon on call for the day — please join them at the bedside?

I dropped everything and ran up to the ward to find an alarmingly dramatic but familiar scene. The curtains were drawn around all the beds bar one, where a group of more than 12 doctors, nurses and other professionals surrounded a patient in the bed. The ECG monitor on the wall showed the heart had stopped. One of the nurses was administering vigorous heart massage, with his hands on the patient's chest, compressing and releasing about 100 times a minute. The anaesthetist was blowing 100 per cent oxygen through a tube he had already placed in the airway. Others were busying themselves with a defibrillator, drugs, monitors and other pieces of kit essential to a cardiac arrest situation. Hospital beds, like most other beds, are not ideal for heart massage, as the mattresses are too soft and give readily under pressure, so the entire mattress was moving and flexing with every compression.

So what was the story? He is a 57-year-old man with known aortic valve stenosis (narrowing of the aortic valve, which is the one that sits at the exit of the heart and through which all the blood to the body must pass). He had been well until recently and nobody knew that he actually had anything wrong with his heart. In the last two weeks, however, he had become increasingly breathless and, a few days ago, he had been admitted to a nearby hospital where the diagnosis of aortic stenosis was made. He was referred to Papworth so that we could complete his investigations, look at his coronary arteries with an angiogram just in case they too needed attention, and then replace his aortic valve, and he was sent home with this plan in place. Unfortunately, yesterday he began to have angina and this rapidly got worse, so he was readmitted to his local hospital as an emergency. They rushed him to Papworth in a blue-light ambulance this morning and a few minutes after arrival in the ward, his heart had stopped.

If you have aortic stenosis, you're pretty safe provided you are feeling OK, because it means the heart is coping well with the narrowed valve. When you become breathless with angina and dizzy spells, it means that your heart is reaching the end of its tether and can no longer generate the force needed to push the blood through the narrowed valve. When that happens, you are at a real risk of sudden death, so something should be done about it and the valve needs to be replaced quite soon if you're to have any form of long life.

This was, in fact, a sudden death from aortic stenosis, which happened on a hospital ward. Had it happened outside the hospital it would be all over by now and the next event would be the planning of his funeral. But it happened here and we were doing our utmost to get his heart started again. Trying to start a stopped heart in aortic stenosis is often not successful, because the obstructed valve is still there, and if it was so bad as to lead to the heart stoppage, it's difficult to see how or why the heart could conceivably start again after all the abuse it sustained during the stoppage and with the offending and obstructive valve still in place. But it's always worth a try.

The cardiac massage continued. The heart normally pumps five litres a minute throughout adult life. When the heart stops, the person doing heart massage tries to pump as much as possible by either compressing the chest to squeeze the heart or, if the chest is open, by squeezing the heart directly with one or two hands. It is exhausting work and that makes one realise just how amazing an organ the heart is — to do this amount of pumping, day in and day out, for a lifetime.

The first nurse massaging the heart was understandably hot, tired and sweaty after 10 minutes or so, and a second nurse had taken over to give him a breather. Unfortunately, but not that surprisingly, the resuscitation simply didn't work and the heart remained resolutely still. Twenty minutes had already passed without a scintilla of recovery in heart function. In a general hospital, the designated leader of the cardiac arrest team would now pause to ask: 'Does anybody here feel there is any point in carrying on with resuscitation?' A few might stay silent, but most of the team members would either say no or simply shake their heads. The leader would then state the time of death and the resuscitation would stop. Someone would go to inform the family and others would prepare the body for transfer to the mortuary.

But Papworth is not a general hospital, and we have equipment and expertise that is not available in a general hospital. One such piece of equipment is the ECMO machine. ECMO stands for 'extra-corporeal membrane oxygenation' and it is essentially a compact and movable heart-lung machine. It can prolong life in someone with a stopped heart, stopped lungs or both.

What we do is push two large pipes into the body, one into a big vein and one into a big artery. Then we connect these pipes to a pump and an oxygenator. We start the machine and the blue blood from the vein is sucked into the machine, where it is oxygenated and turns bright red, then pumped into the artery to supply the brain and the body with oxygen and goodness. The heart and lungs can thus be temporarily replaced, and the patient kept alive for a short time until whatever has caused the heart to stop can be fixed.

The patient was young. His heart problem was amenable to surgery. So we sent for the ECMO machine. We have intensive care doctors who have now become very slick at setting up the machine and connecting it, even in desperate situations like a cardiac arrest. One of them scrubbed up, as did one of the surgical registrars. Between them, they managed to get the pipes placed and connected within minutes — not an easy task while someone is violently compressing the chest at a rate of 100 times per minute, with the chest, the patient and the bed moving up and down like a trampoline.

As soon as the pipes were connected, the ECMO machine started running. Circulation was restored. Heart massage was stopped. The patient was moved to the intensive care unit and we started making the plans for replacing his aortic valve in the next day or two, once his body had recovered from the ravages of a heart that had stopped for about half an hour. There was even a reasonably good chance that he would survive all of this. We'll come back to him later.

Cardiac arrest is the worst medical condition to have, bar none. It is, after all, death. Every acute hospital has a cardiac arrest or 'crash' team and everybody working in an intensive care setting is capable of conducting resuscitation after cardiac arrest. The survival rate afterwards is low, with fewer than one in four patients surviving to leave hospital, but one in four is better than none. In specialist heart surgery units, the survival rate is better at about one in two, so it is truly worth the effort. The reason for that is a combination of two factors: the first is that patients who have a cardiac arrest after a heart operation may have a treatable cause, such as a complication of the operation that can be fixed with another operation. The second is that heart surgery units have all the equipment, staff and expertise ready at hand 24 hours a day and we therefore are well placed to secure the best chance possible of getting a stopped heart beating again.

Sometimes, when cardiac arrest occurs unexpectedly in one of our ward patients after a heart operation, we only resuscitate for a few minutes and if there is not an instant and good recovery of the heart, we simply 'scoop and run'. This means that we immediately transfer the patient to the operating theatre (while continuing the cardiac massage en route), reopen the chest, put in the pipes and connect them to the heart-lung machine. This gets the circulation going again and gives us (and the patient) a breather and some desperately needed time while we look to see why the cardiac arrest happened, and, if we find the cause, try to fix it surgically. We have a dozen or so scoop-and-run events at Papworth every year, and about half of them survive and make it out of hospital. Again, one in two is better than none.

Sometimes a cardiac arrest happens in the ICU, a few hours after a heart operation. If there is no response to immediate resuscitation and we can scoop and run to an operating theatre, we usually do. Sometimes it is not possible to do so, either because the theatres are all full during the usual working day, or because they are empty and all the staff have gone home. When that happens, we don't wait. We simply open the chest on the ICU.

Many years ago I operated on a Norfolk gentleman farmer, carrying out coronary artery bypass grafting for his troublesome angina. His only risk factor was his advanced age of 80. It was a smooth and easy operation and nobody had expected any trouble during the post-operative recovery, but in the middle of the night I received the dreaded call from the ICU nurse, meaning that the registrar was too busy to come to the phone. The patient had had a cardiac arrest, had not responded to resuscitation and the registrar, Reza, was rapidly opening the chest on the bed as we spoke. I asked for the theatre team to be mobilised and sped back to the hospital.

When I arrived, Reza had already opened the chest and was massaging the heart internally, squeezing and releasing the heart in his right hand to keep some sort of blood flow going. It had taken me 30 minutes to get dressed and drive to the hospital, and Reza's right hand was getting very tired, so I took over. It took another half an hour before the theatre team arrived, and longer still before they had set up the theatre equipment. In the meantime, my hand was becoming extremely tired and this was not helped by the on-call anaesthetist who was grumbling and constantly making comments like 'This is an utter waste of time', 'He's 80 for goodness's sake' and 'He'll never leave the hospital alive', and so forth.

We got to theatre and put him on the bypass machine, much to the relief of the exhausted muscles in Reza's right hand and mine. Within minutes the heart started again. We checked all the bypass grafts and they were all fine, so we shrugged our shoulders and closed the chest. The patient went home 10 days later and, 10 years after that, I attended his ninetieth birthday party.

It takes a lot to improve the results of heart surgery nowadays. Overall, the death rate from open heart operations in the UK is hovering around 2 per cent, and this includes many urgent and complex operations done on extremely sick and old people. We have come a very long way in improving and refining the techniques used in the operating theatre and we have learnt how to look after the heart well, while we subject it to our surgical assault. We have also worked quite hard to refine our systems, to reduce the possibility of human error both inside and outside the operating theatre. Despite all of this, nasty horrible things can still happen to patients who have had heart surgery, and the next improvement may well be in the hospital's ability and willingness to salvage patients from the jaws of death when such things happen. I believe that one of the reasons we do so well at Papworth is that we do have the ability and willingness to go that extra mile and to pull patients back from the brink when such disasters as cardiac arrest, rare as they may be, inevitably take place.

Cardiac arrest is truly appalling when it happens out of hospital, at home, on the road or in a public venue. Even if a bystander witnesses the arrest, has a reasonable idea of how to conduct resuscitation and cardiac massage, and begins to implement that immediately while waiting for the ambulance, only 1 in 10 such people will leave hospital alive, and this takes me to John Keegan.

I am not here talking about the eminent military historian, but about my next-door neighbour who ran a company which manufactured industrial catering machines. I used to tease him by referring to his work as 'the sausage-making business'. A lovely and sociable man, with a well-developed sense of humour and an intact inner child at the age of 60, he used to play cricket, rollerblade round his drive and build model airplanes. My younger son Ramsay would often knock on his door at weekends to ask Maureen, John's wife, if John was 'allowed to come out to play'. John also smoked about 80 cigarettes a day, frequently lighting one cigarette from the butt end of the previous one.

One evening, after a long day's operating, I settled with a beer in front of the television to watch the evening news. Suddenly, there appeared to be a commotion with a lot of noise coming from outside the house, with shouting and wailing, so I reluctantly got up to investigate. I opened the back door to find Maureen wandering around our shared driveway in a state of near hysteria.

'It's John,' she said. 'He doesn't look right.'

I walked across the drive to their house and there he was, in the kitchen, slumped over the table with a stopped heart. I tried to find out from Maureen how long he had been in this state, but she was too distraught to say with any certainty.

I had to make a quick decision between doing nothing and letting him die or starting resuscitation, and, if successful, possibly risk ending up with John alive, but with serious brain damage, because the brain will not survive a stopped heart for more than a few minutes. It was a difficult call with very little time available to make the decision. I decided to give him a chance.

I shouted to Maureen, who was still wandering about in a state of agitated distress, to call for an ambulance, give her address and specifically to use the words 'cardiac arrest' when speaking to the operator. I then moved John off the table on to the floor and began the chest compressions for cardiac massage. The kitchen television was still on and was blaring a particularly intrusive set of hard-sell commercials. I was pondering whether it would be considered ethical to stop the cardiac massage for a short while to switch off the TV, since the remote control was not within reach. Then an exceptionally irritating advertisement for a toilet cleaner came on and I abandoned the resuscitation briefly, got off the floor, went to the TV, silenced it and resumed cardiac massage.

Maureen came back into the room, slightly calmer now. She had made the call and the ambulance was on its way. I asked her to go outside and stand at the end of the driveway, and, as soon as the ambulance arrives, to ask the paramedics to bring the defibrillator with them and guide them to the kitchen. Some 15 minutes later, while still doing the chest compressions on the kitchen floor, I saw the reflection of the blue flashing lights through the door to the hall and two paramedics walked in carrying the defibrillator.

'He's in cardiac arrest,' I said. 'His only chance is if it is ventricular fibrillation, so let's connect the defibrillator now and shock him if it is.'

'And who the hell are you?' said one of the paramedics.

'Hang on,' said the other, 'I know you,' and to his colleague: 'That's Mr Nashef. He did that aortic dissection I took to Papworth last week.'

We connected the defibrillator. The cardiac arrest was in fact ventricular fibrillation, the type that can be treated by an electric shock. We administered one shock and the heart restarted instantly. Within seconds, John had started to breathe again. They were short, gasping and irregular breaths, but where there's breath, there's life. We moved him on a stretcher to the ambulance, which sped towards the coronary care unit at the local hospital, only a mile or so away.

An hour later I thought I had better go to find out how he was, and I was still very worried that he might have sustained brain damage. I drove to the hospital and walked into the coronary care unit to be greeted with a loud shout: 'You bugger! I think you've broken half my bloody ribs! But I knew that one day it would be useful to live next door to you!'

John was sitting up in bed and smiling. Later, he was investigated at Papworth and found to have just the one coronary artery narrowing in a branch of a branch of the left coronary artery. The branch in question was of very little significance and his heart had continued to work well despite the blockage, but the tiny heart attack he had sustained had caused a near-fatal but temporary disturbance of his heart rhythm.

Somehow the tens of thousands of cigarettes he had consumed had had only a very minor effect on John's coronary arteries. Nevertheless, the consequences of that tiny blockage were a wake-up call. This was enough for John to stop smoking completely, sell the business and retire to the north of England. Before he did so, he tried to claim on his life insurance. He telephoned the insurance company and said to the hapless insurance claims handler: 'I'm calling to claim on my life insurance, having died a few weeks ago.' The claims handler did not know quite how to handle this particular claim and referred the matter to his manager, who was smart enough to demand a death certificate. Of course, John was not in a position to provide that document, and the claim was not successful.

And what about the 57-year-old man on ECMO? He is now at home. After two days on ECMO, he had his aortic valve replaced by my colleague Steve Large and a couple of bypass grafts thrown in. He went on to make a slow but steady progress back to life from having been, for over half an hour anyway, well and truly in the land of the dead.

CHAPTER 18

An accidental fraud

A few years ago I was consulted by a man in his late eighties. More than two decades previously he had received a coronary artery bypass grafting operation in another hospital. The bypass grafts were working fine, but over the years he had, as so many older people do, developed a tightly narrowed aortic valve and his coronary artery disease had progressed somewhat, so that another coronary artery, which was not touched at the first operation, was now blocked. He was keen to explore the possibility of a second operation and had asked his general practitioner to refer him back to the hospital where the first operation was carried out. He was turned down for surgery there because it was not felt that the benefits could justify the risk at his advanced age. He tried two other hospitals with the same disappointing result and finally presented himself at Papworth.

When I saw him, he was a slim, sprightly and energetic

old man who still had all his wits about him. I listened to his complaint of breathlessness and explained to him that the risk of a second-time, double-procedure operation would be substantial and that, at his age, it would be unlikely that such an operation could actually prolong his life. In fact, an operation could actually bring his life to a most abrupt end, so that I wouldn't dream of considering it as an option for him unless his symptoms were very severe, producing a substantial limitation on the quality of his life.

I then questioned him in detail about the nature and severity of his symptoms and he declared again that his only problem was that he became short of breath when he exerted himself. I asked him how much he could do before breathlessness stopped him and he said that if he tried to walk more than a couple of miles at a fast pace or uphill he became breathless. I then told him that there were many folk in their eighties who would be more than happy to be able to walk fast for two miles and that I did not consider this to be severely limiting, and he replied, 'You don't understand: I'm a hillwalker. I live for hillwalking. If I can't climb hills, I don't really want to live. I don't care tuppence about the risk. Now are you going to fix me or not?'

Of course, after that, he got his operation. He spent a few days longer in hospital, because of his age and the complexity of the procedure, and went home in good shape. A few weeks later I received a postcard from him. It featured a photograph of him standing at the top of a hill in Derbyshire with a brief note expressing his gratitude. That postcard made my day and is precisely the sort of thing that makes it all worthwhile.

I sometimes give career guidance to Cambridge medical students as part of a surgical specialty 'fair'. In this sort of event, representatives of the various strands of surgery lay out their wares and give short talks to the students in the hope of attracting and recruiting the brightest and the best to their branch of the profession. When it is my turn to talk about heart surgery, I do my best to tell it like it is. I explode a few myths, such as heart surgery is ridiculously competitive and impossible to get into, because it's not. It is probably more competitive than some specialties, but it is accessible and achievable for any aspiring doctor prepared to put in the work and the hours. I also debunk the myth that it is a dying specialty because, allegedly, soon the cardiologists will be able to do with a catheter and an X-ray machine everything we surgeons do with a scalpel and an operating theatre: they will not. The cardiologists may have already creamed off some of the easier procedures, but in the end it is often only an open heart operation that will fix the big breakages in the heart. I still recall that the first time I heard a cardiologist declare heart surgery to be a 'dying specialty' was over 20 years ago. During that time, heart surgery has continually expanded in volume and in the range of treatments that it offers, so that, to paraphrase Mark Twain, the reports of its death have been greatly exaggerated. I also put paid to the notion that you must have phenomenal manual dexterity to succeed as a heart surgeon: you do not. Anyone who is not all thumbs can be taught the required skills. I am sometimes asked if I have had my hands insured. Of course I have not, but if I were to buy an insurance policy for any body part essential to my career, it would be for my brain rather than for my hands.

After this somewhat encouraging busting of myths, I tell them some discouraging stuff that they should also hear in the interests of balance: heart surgeons are not rich. They are not exactly poor either and will not be struggling to feed their families, but their earnings are nowhere near what some other surgical and medical professionals can achieve. It may be a glamorous profession to declare on meeting a stranger in a pub, but real money in surgery flows in specialties where several short and easy operations can be performed in one sitting. By comparison, a heart surgeon's standard working day of two four-hour-or-longer operations is not likely to turn into a money-spinner. I also tell them about the endless commitment that heart surgeons feel for their patients, and the heartache that complications bring, together with the huge disruption of social and family life when these complications happen after hours. In heart surgery, the responsibility for a patient does not end when the surgeon gets home, regardless of whether that surgeon is on call or not. We are always on call when it is one of our own patients.

Another essential feature of heart surgery is one that does not often spring to mind: it is rarely, if ever, '*ablative*'. Many surgical specialties deal with a problem by removing the bit of the body where the problem lies. In fact, a well-known maxim of general surgery is 'If in doubt, cut it out!' This is more prevalent than you think, and goes way beyond the obvious example of amputation for gangrene. Think about the appendix, the gall bladder and varicose veins, which are the fodder of a huge number of operations in which the offending part is simply removed. Add to that all of the operations on toenails and lumps and bumps, plus almost every operation done for cancer. In fact, when I was in training, the standard operation for a benign stomach ulcer was to cut out part of the stomach. If the ulcer was in the duodenum, the surgeons, on finding that the duodenum could not be easily cut out, went ahead and cut out part of the stomach anyway! (This was ostensibly to reduce the acid production that was blamed for the ulcer, but now we know things are a lot more complicated than that and we have medicines that can treat ulcers better than any surgery achieved in the past.) General surgery, neurosurgery and large chunks of other surgical specialties are still substantially ablative, whereas heart surgery simply cannot be: you can't cut out the heart (unless it is to replace it with something), so you simply have to find ways of fixing it by bypassing blockages, repairing or replacing faulty valves and so forth. This *reconstructive* nature of the specialty is one that I (and many others) find so appealing.

Where heart surgery truly excels is in two vital areas. The first is that there is not a single specialty in medicine that does more good. The overwhelming majority of our patients find that they feel hugely better after an operation. The symptom that took them to the doctor in the first place will have gone. Angina disappears completely in over 90 per cent of patients who have a heart operation. Breathing improves measurably in a similarly large proportion. Patients regain a normal quality of life and their sick hearts no longer restrict them or their activities, but that is not all. If heart surgery is effective when it comes to helping people feel better, it is even more so when it comes to helping people live longer. The majority of heart conditions that we treat are not just awful to live with, they also shorten life. A successful heart operation will result in substantially longer life for many, whereas an unsuccessful one will result in an abruptly shortened life for the few. The net result is a huge number of net life-years gained, so it is a field which yields generously both in quality of life and quantity of life, and, in this context, there is no medical field that comes close. Because of this, I don't think that I have ever gone home after a long day's operating thinking, 'Well, that was a waste of time.'

The second area is that both the rewards and punishments that this specialty dishes out to its practitioners are immediate and they are directly related to how well the operations went. By and large, barring a few exceptions and bolts from the blue, and assuming a good level of decision-making in selecting an operation for a particular patient, the connection between how technically good an operation is and how well the patient responds and recovers is a very tight one. The aphorism 'Cut well, sew well, get well' is largely true in this specialty.

All of this means that heart surgery is a job where the rewards, even if not financially lavish, are rich and plentiful in so many other ways. What's not to like? My now retired colleague John Wallwork was fond of repeating the mantra that we heart surgeons are so very fortunate in that our job is one of the very rare professions where we are actually 'paid to have fun'. I am not sure that I fully agree with him that 'having fun' is the best way of describing what we do, but I do not for a minute doubt that there is a huge enjoyment factor and that the satisfaction of doing a lot of good for a lot of people is hard to beat.

During my career, the number of surgical trainees whom I have encountered on our training programme over the decades must be in the hundreds. Some were born to do heart surgery and some had it thrust upon them. Others have rationally and systematically explored all the medical and surgical specialties before making an informed decision that heart surgery was, in fact, their career of choice. These trainees, and the medical students whom I teach regularly, have often asked me when and how and why I decided to be a heart surgeon. Others have wondered about the sequence of events that led a Palestinian boy who was raised in Beirut to end up as a heart surgeon in Cambridge. I was most certainly not born with that calling, nor did I ever make an informed decision based on facts. In reality, I became a heart surgeon through a series of serendipitous accidents and stupid decisions made in absolute ignorance.

Most of my family had studied literature, education and

philosophy and there was not a single doctor among them. As I was growing up in Beirut, I was reasonably good at science in secondary school, but had no idea what to study at university other than something vaguely related to science, such as engineering. I was about to go to an A-level college in Loughborough to study physics, mathematics and applied mathematics, since that seemed to offer a decent grounding for such a career aspiration, vague as it was, when I was struck with an unusual pneumonia. After I had suffered a few nights of high fever and hallucinations, my parents and our family doctor were getting worried and had me admitted to hospital.

The hospital in question was the American University Hospital in Beirut, a gleaming, modern building enviably located next to the university campus and overlooking the Mediterranean. I was admitted to a private room on the ninth floor with a fantastic sea view through the floor-to-ceiling glass window. Within a couple of hours of arriving my fever had taken its course and I felt much better, but the attending physician kept me in hospital for four more days while the lung changes on my chest X-ray improved, and so, I spent four healthy days feeling quite well and observing the hospital environment.

The nurse who attended me was beautiful, bubbly and blonde. Groups of medical students came to visit me and to listen to my chest and they were charming, attractive and very happy people. I must have been the only patient of their generation on the floor as they seemed to choose to take their coffee and lunch breaks in my room. We talked a lot and almost became friends and I could not help thinking that here was a group of young people who seemed massively to enjoy what they were doing and in whose company I felt very comfortable. I also thought that they looked super cool in their white coats and, on top of all that, I'm ashamed to say, I really enjoyed playing with the buttons which moved, reclined and adjusted the elaborate and hi-tech electrically powered hospital bed. That, and the very large colour television set with a remote control (a novelty in those days) clinched it for me. Forget engineering: I decided to become a doctor, and the decision was entirely based on one pretty nurse, some cool-looking medical students, an electrically operated bed with a sea view and a television set.

I arrived in Loughborough and immediately switched from applied mathematics to chemistry, so as to fulfil medical school requirements. I shared a tiny room in 'digs' with another A level student at the college. He was from Hong Kong and spent most of his time in London playing the stock market, so that he rarely turned up either for classes or in our digs, which was a relief as the room was barely big enough for one.

More importantly, there was nowhere to store food and no cooking facilities. Eating out rapidly became economically untenable, as a simple calculation showed that I would fritter away my monthly allowance in a week if I relied solely on fish and chips, and even more quickly if I varied the diet. I had to think of a way to eat cheaply at home and decided that some bread and olive oil, the poor Arab's staple food, would have to do. I went to several shops and supermarkets to try to find some olive oil and was met by blank stares from all the shop assistants. Finally, one of them pointed out a large store across the street and said, 'I think they may have some in there.' I crossed the street and walked into said shop, which immediately struck me as an extremely unlikely place to stock any olive oil. I nevertheless asked if they had any and was directed to a particular aisle. Sure enough, there it was: an exorbitantly priced tiny bottle, containing a small amount of pallid olive oil that would be barely enough for drizzling on a single salad, with instructions on the label on how to deliver the oil up one's bottom. The shop was Boots. Nowadays, supermarket shelves in the UK groan under the weight of dozens of beautiful and characterful olive oils from all over the world, and seeing the range, quality and variety of what's on offer always reminds me of that visit to Boots all those years ago.

In college, my fellow classmates were a motley crew. Some had transferred from mediocre schools in the hope of achieving better A level results in Loughborough College, which focused on sixth-form studies among a few other things. Others were repeating previously unsatisfactory A level attempts, or, like me, had come from abroad with the specific aim of acquiring the A levels that would allow them to attempt entry into a British university.

At that time, medical school entry was highly exclusive and unquestionably traditional. The typical successful applicant to medical school was an upper-class white male, educated at a reputable public school and preferably the son of a doctor who qualified at the very medical school the applicant was hoping to join. I do not exaggerate: these were the features that often decided who succeeded and who did not. However, I knew nothing of all this and had naively assumed that medical school entry would be purely meritocratic.

One morning, the headmaster came into our class, just one of dozens of A level classes in the college.

'Who here wants to study medicine at university?' he asked.

More than half the class put up their hands.

'I have something important to tell you,' he said. 'This college would be considered very lucky — no: extremely lucky — if one, just one, of our many A level students in the entire year group made it as far as being interviewed for a single medical student place this year. In other words, the chances of any of you getting an interview are smaller than 1 per cent. Think again, and please choose something other than medicine.'

Having dropped that bombshell, he walked out.

I duly filled in the application form for university entrance, placed four universities as my top choices for medicine and added a bottom back-up choice for engineering, as instructed by the head teacher. That year, the college was indeed 'lucky': as I was that one student who was granted an interview for medical school. I turned up at Southampton

University, utterly ignorant of the system, its traditions and requirements. I wore a wholly inappropriate snazzy brown spiv suit with a ridiculously wide psychedelic tie. I had shoulder-length hair and looked more like an applicant for a job as a nightclub disc jockey than a serious and conservative future doctor. My interviewers asked many questions about why I wanted to be a doctor and tried to find out if I knew what a medical education was about. It was immediately apparent that I knew little about medicine, even less about medical education and absolutely nothing about the NHS. Needless to say, I was not offered a place at Southampton. I returned to Beirut and entered the American University there, the very institution whose shiny, gleaming hospital had converted me towards a medical career when I was briefly an in-patient there three years previously.

In Beirut the university followed the American system and medicine was a postgraduate course. Aspiring medical students had to do a bachelor's degree with a core 'pre-med' curriculum, and then compete for medical school entry based on their performance in this undergraduate work. On top of that, the university had a unique feature: all undergraduate students, including us 'pre-meds', were required to take four semesters of Cultural Studies, and this mysterious course began with a lecture on a Tuesday. Since the entire freshman year was there, hundreds of us filled the largest lecture theatre on campus and waited for the lecture to begin, somewhat unsure what Cultural Studies meant.

A Swiss professor named Heinrich Ryffel walked onto

the podium. He wore a very crumpled linen suit and was smoking a pungent pipe. He held up a book called *The Epic of Gilgamesh* and lectured us about it. It was the first literary work ever, written more than 4,000 years ago. He spoke about its provenance in Mesopotamia, highlighted its importance and tried to convey the essential messages that the book carried. Then he instructed us to read it immediately. On Thursday we had a discussion seminar about it. On Monday we were told to write an essay about it. Then on Tuesday we all trooped in to the same massive lecture theatre for the next instalment: the pre-Socratic philosophers. Again: read, discuss, write the essay. Then Socrates the following week. Then Plato. Then Aristotle. Then the Old Testament, the New Testament, the Qur'an, St Augustine, the Renaissance works, Kant, Hegel, Hobbes, Mill, Rousseau, Descartes, and all the way through every single important landmark of human culture until we reached the end of the course and the very last tome, which was the Little Red Book entitled *The Thoughts of Chairman Mao*. By the end of the two-year course, I had not only secured a place at medical school, I had received a true education.

I started medicine the following year, and the Lebanese Civil War broke out. The country split along several sectarian and political divides, which were fluid and unpredictable, so that at one point or another every sect had aligned itself with almost every other sect in fighting every other sect. What this meant was violence and murder on a grand scale, roadblocks which sprang up out of nowhere and at which innocent civilians were summarily shot if they were perceived to be 'the enemy' at that particular point in the war. The city of Beirut was divided along sectarian lines and the battle was raging only a few blocks from the medical school. There were food and fuel shortages, power cuts and interruptions to the water supply. There were snipers who took potshots seemingly at random, and my car had two bullet holes to prove it.

Throughout this mayhem, my overriding thought was not fear of the civil war itself – it is remarkable how quickly one can adapt to new circumstances and somehow manage to live with them. My main concern was a simpler one: it had taken me three years of hard graft and stiff competition to secure a medical school place, and now it looked like my medical school was being threatened with disruption and likely closure due to the war. I started to apply for a transfer elsewhere. I applied to hundreds of universities, in any country where I could speak the language. I was rewarded with hundreds of rejection letters which I used to wallpaper my bedroom. Finally, two universities said yes: Montpelier in France and Bristol in England. Montpelier wanted me to start at zero, whereas Bristol recognised my first year in Beirut and accepted me into year two. Bristol it was. And the American University of Beirut? Well, despite the raging, bloody war on its doorstep, it never did close its doors. My year-group friends made it, and most of them became very successful doctors.

I arrived in Bristol in the beautifully hot summer of

1976 and joined the second year of the preclinical course. After completing that year of lectures, laboratory work and microscopes, we started year three in which we would finally be allowed to see real, live patients. I was with the group of medical students assigned to Ward 20 of the Bristol Royal Infirmary. At long last, the time had come to play doctor properly.

Feeling very cool in a crisply starched white coat, I entered the ward. It was a traditional NHS Nightingale ward: a huge and grey dormitory with two long rows of beds. The few windows let in the grey light of the rainy September day outside. There was no gleaming white room overlooking the Mediterranean, no electrically powered beds and no colour televisions. The patients were anything but glamorous. The nurses looked harassed. The place smelt of stale urine. My hastily made and ill-conceived decision to study medicine suddenly seemed like the biggest mistake ever. Was this ghastly environment now going to be the backdrop to my life? What had I done?

The second accident was in deciding to become a surgeon and this was for even stupider reasons. I was in London as a surgical 'houseman' or house officer, the lowest rank of the hospital medical team. Among my many, chiefly administrative chores was to ask physicians for advice about surgical patients with complex medical problems. Since I then wanted to be a physician, I resented having to ask outsiders to help with medical problems that I thought we could handle ourselves, but my surgeon bosses were not interested in these problems: they just wanted to operate and leave the finer points of drugs and potions to the experts. What was worse, however, was that whenever we asked a medical team to review a surgical patient, it was my duty as a houseman to accompany them on their visit, answer questions about the patient and implement their recommended treatment.

The problem was that, as I saw it, these medical teams seemed to have a lot of time on their hands. They would descend en masse on our surgical ward: professors, lecturers, consultants, trainees, fellows, visitors, house officers and medical students. The entire mob would surround the hapless patient (and the even more hapless surgical houseman) and ask many questions. Then they would begin to discuss the most recent research, vying with each other to quote the latest hot-off-the-press offering in this or the other journal. As a busy surgical houseman, I had very little spare time on my hands and the last thing I needed was to waste any of it listening to their irrelevant and lengthy verbosity, when there were dozens of clinical jobs clamouring to be urgently done on the hectic surgical wards.

It seemed that the hottest topic for physicianly debate back then was whether potassium supplements should be given to patients receiving diuretic drugs, as such drugs are known to cause loss of potassium. From my point of view, the debate appeared Byzantine. Who cares? Give potassium routinely, or don't give it, or give it only if the potassium level drops: they all seemed valid options and hardly the sort of stuff to get excited about, but for a few weeks, every time

I asked for a physician's opinion the horde descended and started to argue about this mindlessly tedious topic. I began to dislike physicians intently.

One day I was asked by my surgical bosses to request a review of one of our patients with Parkinson's disease. As I did so, I told myself that if they mentioned potassium just one more time I no longer wanted a career as a physician. The mob came and assembled around the bed. To the best of my ability I pointedly kept the discussion focused on Parkinson's disease, and they were just teetering on the brink of telling us how to manage it when one of them, a lecturer, picked up the drug chart.

'I see the patient is taking a diuretic,' he said.

'Ah!' said the professor, 'how interesting! Is he receiving potassium supplements? Because this morning's *Lancet* has a paper which shows that … '

I had already stopped listening and resolved to become a surgeon.

I had no idea what surgical specialty to go for, other than a vague inclination to avoid any career choice which would entail a lifetime of peering down, through and up orifices. This excluded ear, nose and throat surgery, urology, gynaecology and proctology. I kept an open mind for other specialties until a couple of years later when I was doing my basic surgical training in Exeter and I worked for Mike Pagliero, a charming thoracic surgeon who specialised in lung and gullet surgery. He allowed me to do a fair bit of operating, despite my relative inexperience. I thought I had decided to become a thoracic surgeon because I liked the specialty, whereas, in truth, I just liked Mike Pagliero. I told him my decision and he was very supportive, but he informed me that in the UK there is no such thing as a dedicated thoracic surgery training programme: thoracic surgery was part of cardiothoracic surgery and I would have to train in heart surgery as well, and then ditch the heart bit to concentrate on lungs and gullets much later. Fine, I said. How hard could the heart be?

Deciding to go for surgery as a career meant that I had to become a Fellow of the Royal College of Surgeons (FRCS). In those days, the Fellowship was obtained by taking an examination in two parts. The primary part began with a multiple-choice paper in only three theoretical subjects: anatomy, physiology and pathology, and those candidates who achieved a reasonably good score went on to a viva examination, where they were asked more questions about the same three subjects, using dissected cadavers and dusty and archaic pathology specimen jars. Patients and surgical topics formed no part of the primary examination: it was all basic (and, for an aspiring surgeon, bloody boring) science.

I was working on a (now illegal) rota, which precluded any form of serious study: on call one-in-two. This meant non-stop work (with occasional snatched short naps) from Monday morning to Tuesday evening, Wednesday morning to Thursday evening, Friday morning to the following Monday evening, Tuesday morning to Wednesday evening and Thursday morning to Friday evening. After this arduous stretch, there finally comes one brief weekend off duty every fortnight. Had my colleagues and I been sensible, we would have ideally devoted this valuable weekend to studying for the Fellowship exam and to catching up on the lost sleep from the previous two weeks. We were not sensible and did neither of these things. Instead, for most of us junior hospital doctors, that priceless weekend after two weeks of working hard was instead dedicated to partying hard, which meant that we started the next gruelling fortnight as tired and sleep-deprived as we finished the last one.

I had done virtually no studying when I took the primary Fellowship exam and, of course, failed it. Twice. I did not even manage to get past the multiple-choice section either time. The third time I scraped through (just) and went on to the viva. I struggled with the questions, but did the best I could with my limited knowledge, trying hard to convey an impression that I was, actually, somewhat familiar with the anatomy of the human body. My examiner then took me to a dissected cadaver and pointed to a nerve that coursed between the lung and the heart on its way to the diaphragm. He asked me what nerve it was. It was a relatively easy question and I replied correctly that it was the phrenic nerve. He then asked me if I knew what the word meant. My father had studied classics and had tried to teach me some philosophy and ancient Greek, in which *phrenos* referred to the mind, so I again replied correctly. He asked why the phrenic nerve should be thus called, when it controlled the breathing muscle that is the diaphragm. I replied by wondering out loud if the Greeks felt this was the nerve by which the mind controlled the guts, and manoeuvred the conversation towards Plato's divisions of the personality and of society. My examiner was visibly impressed. We had a brief but animated conversation about Greek philosophy and for the rest of the exam he asked me ridiculously easy anatomy questions along the lines of 'What's that?' 'The liver, sir', 'Jolly good. Have you actually read Plato's *Republic*?' 'Yes, sir', and, from that point onwards, I could simply do no wrong. I passed the examination, not by having expert knowledge of the subject matter, but by having a little grounding in the Classics (thank you for that, Dad, and my Beirut university) and by being lucky enough to meet an examiner who appreciated this utterly irrelevant erudition.

In those days what mattered far, far more than knowledge was whether 'your face fits'. Tall, handsome, public-school educated rugby players from the Home Counties tended to pass the exam with inadequate preparation, whereas short, unsporting, comprehensive school products with a background from the Indian subcontinent could fail, despite having all the requisite knowledge. Women especially had a hard time entering this far-from-meritocratic profession. Things have changed, and today's examinations are conducted in a much fairer and well-governed manner, with integral processes designed to prevent unfair discrimination in the profession. Back then, however, whether your face fit played an important role and, for some strange reason, mine luckily did.

My knowledge of anatomy was appalling when I passed the primary Fellowship and is still, in fact, embarrassingly patchy to this day. In my first year in Beirut we had covered the anatomy of the arm, the thorax and the abdomen. When I transferred to Bristol straight into the second year, my fellow students had covered the arm, the leg and the thorax in their first year. This meant that I went through the anatomy course across two medical schools, but without ever looking at the anatomy of the leg. A few times during the medical course I idly mused that I really must, someday, at least take a glance at the anatomy of the leg and try to learn the basics, but other, far more pressing demands on my time always took precedence. These included clinical studies as well as parties, beer, crosswords and the rock band in which I played guitar (badly).

A few years later I was a casualty officer in a London teaching hospital when a woman presented with a sore ankle, having twisted it badly on the edge of a pavement. I examined her and she had all the classic hallmarks of a fracture: swelling, bruising, exquisite tenderness and a visible angle where there should definitely not be one. I requested an X-ray, which showed the fracture very clearly, so I immediately bleeped the orthopaedic surgeon on call for the day, very chuffed with myself for having made the correct diagnosis.

'Hello, it's the casualty officer here and I've got a lady with a fracture,' I said, triumphantly.

'Where?' he asked.

'In the Casualty Department, of course.'

'No, I meant where is the fracture?'

'In the left leg.'

'Where in the leg?'

'In the ankle.'

'You don't get it, do you? Now listen carefully to what I am about to ask: WHICH – BONE – IS — BROKEN?'

The orthopaedic surgeon was shouting now, and, just then, it suddenly dawned on me that I had absolutely no idea which bone it was, having never studied the anatomy of the leg. A few seconds of awkward silence ensued.

'Erm, you know how there are two long ones? Well, it's the thinner one of the two that's broken.'

I cannot fully remember the terminology he employed next, but it most certainly included some juicy expletives. Whatever the exact words were, their subtext was definitely along the lines of 'What kind of utter incompetent idiots are we now appointing to the Casualty Department at this hopeless hospital?' — or something to that effect.

About a year after my primary examination I took the final Fellowship examination, again from a one-in-two rota and again with nowhere near the required knowledge to be a surgeon. The 100-or-so candidates who took the final viva exam that day were numbered 200 to 299, and we all gathered around the statue of John Hunter in the august and imposing atrium of the Royal College of Surgeons in London. We were all waiting for the verdict. After a seemingly interminable wait, the college porter finally shuffled out and stood underneath the great statue. We all fell silent. He produced a scrap of paper from his pocket and started reading out the numbers of the successful candidates.

'Two hundred and twenty-seven,' he proclaimed as 26 faces dropped, and 26 despondent figures dragged their sorry legs out of the college. He continued, and with every successful candidate number, a bunch of others exited the college, defeated. He read my number. I had again scraped through, probably because 'my face fits'. Finally, there were only about a dozen of us successful candidates left standing in the college atrium. He instructed us to wait there and walked out. Some 15 minutes later, the members of the College Council appeared in their full regalia. They wore long black ceremonial robes embellished with red and gold and marched slowly and with gravitas past us in a single file procession towards one of the meeting halls. The porter reappeared and motioned to us successful candidates to follow them. We all trooped into the hall and the doors were closed. We were each served a thimbleful of cheap sherry. After some 15 minutes exchanging bland pleasantries, the council members majestically walked out again. I was now a proper surgeon, with an FRCS, and ran out with all the other new proper surgeons with an FRCS to the pub for a proper drink.

And so I became a surgeon through a series of stupid and misguided decisions based on the most spurious motivations and arising from inconsequential and barely relevant events; and after five more years of training in cardiothoracic surgery, I similarly became a heart surgeon after even more ill-advised decisions and seemingly random events, but I am glad to say that I have never regretted that career choice.

Perhaps as a result of the profound ignorance I have had of what I was letting myself in for, every step of the way I felt like a fraud. From becoming a medical student, to a qualified doctor, to a surgeon-in-training, to a heart surgical specialist trainee and all the way to senior consultant surgeon, I always felt, deep down, that I had no right to be there.

This feeling is a well-recognised feature of 'the impostor syndrome' a term coined in 1978 by two American clinical psychologists, Pauline Clance and Suzanne Imes. The syndrome describes the inability of some high-achieving individuals to accept and enjoy their accomplishments, and one of its features is a persistent fear of being exposed as a fraud. I had it, big time. I was constantly in awe of my supremely confident peers, who invariably seemed to know everything and relished sharing that knowledge. They always knew what they wanted in their professional life, had made a firm career plan and behaved as though they were made for the job. In contrast, I was always full of self-doubt, unsure I was worthy of being let loose on innocent patients and certain that I was most definitely not made for the job. I always had the nagging feeling that I was, to some extent, a fraud, but, over the years, that feeling became less and less troublesome. It would be nice to say that this happened because I gained more knowledge and confidence. It did not. What gradually dawned on me over the years was that, after all, those supremely self-confident individuals did not know and never had known everything. They too were, to some extent, frauds, but they were far more adept at hiding it. We doctors never know everything but, fortunately, most of us recognise our limitations, are capable of seeking knowledge when ours is lacking and do not hesitate to marshal the expertise of others when it can help us and our patients, and, as long as we have the humility to accept this fact, we muddle along just fine.

Let us now move on to a different kind of impostor.

If you can dream — and not make dreams your master;
If you can think — and not make thoughts your aim;
If you can meet with Triumph and Disaster
And treat those two impostors just the same

. . . .

If you can fill the unforgiving minute
With sixty seconds' worth of distance run,
Yours is the Earth and everything that's in it,
And — which is more — you'll be a Man, my son!

So wrote the great poet Rudyard Kipling in his poem with the laconic title 'If—', reportedly the British nation's favourite poetical work. In this poem, Kipling advocates the development of elaborate indifference to 'those two impostors' triumph and disaster. He seems to say that we should adopt a stoical, or almost Buddhist, approach when it comes to reacting to triumph and disaster, and that we should learn to ignore them in order to become fully mature and enlightened.

I beg to differ.

Triumph and disaster are precisely the very factors that motivate us, inspire us and drive us forward. We learn from them, we celebrate one and mourn the other, and our professional life, to a large degree, consists of striving endlessly to achieve triumphs, while dodging disasters. In short, they are the stuff that life is made of. Heart surgery teems with both of them with an abundance that no other vocation can match. The elation from a surgical triumph knows no bounds, whereas the misery of a disaster can reach an abysmal profundity.

As my career progressed through the years, the triumphs have continued to accumulate and, since we now operate on older and sicker patients and perform ever more complex and dangerous procedures, both the number and the magnitude of these triumphs have steadily increased with time. What has not increased, however, is the elation that should come with them: on the contrary, such elation is slowly but perceptibly diminishing in intensity. Perhaps this is as a result of the familiarity that builds up with the passage of time and the rising level of expectation; perhaps it is a natural by-product of simply getting on in years, but I no longer get quite the same 'kick' out of a triumph from a hugely complex and dangerous operation. Paradoxically, the misery from the occasional disaster remains unchanged in both magnitude and impact. I imagine that one day, if present trends continue, I may find that triumph elation no longer compensates for disaster misery and then, perhaps for purely selfish reasons, I'll put the scalpel away for good.

But not yet.

APPENDIX:

My crossword puzzle for Roger Whiting

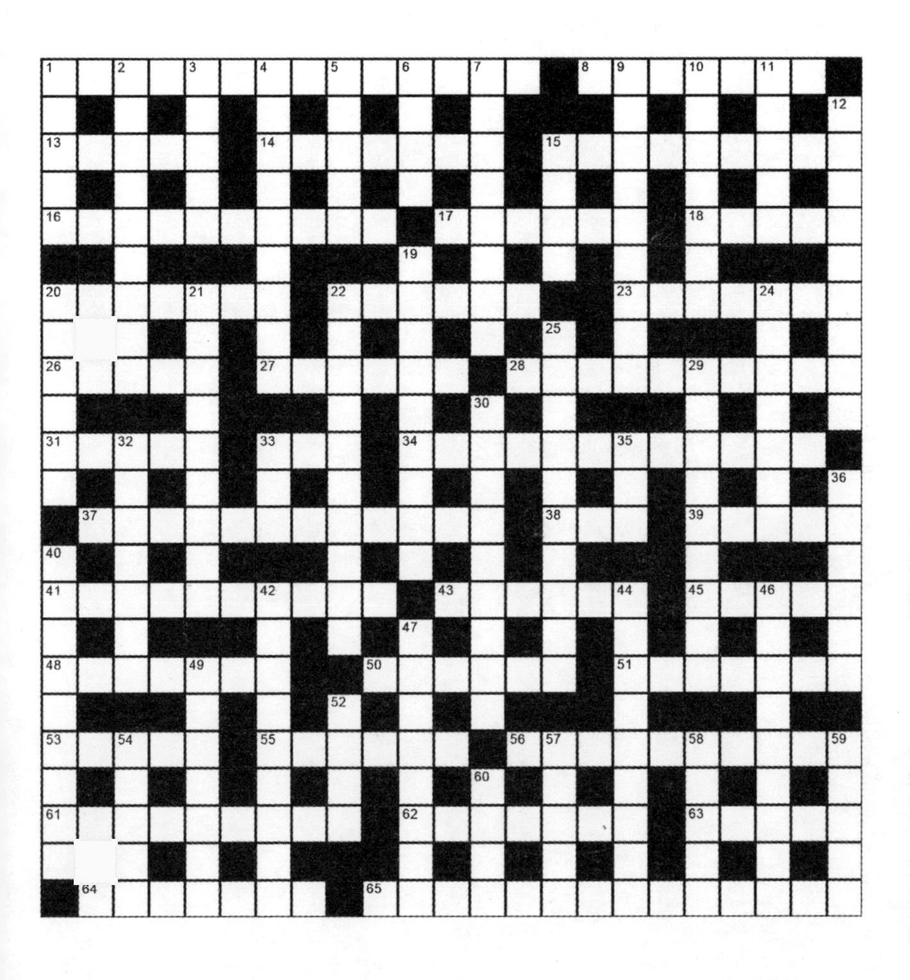

Across

1/8/12 Planning exit, you ask high-tech light-hearted reference book (3,11,5,2,3,6)
13 The muscles of erection (5)
14 Walk back with unfinished clue to Arion's rescuer (7)
15 Game for a dance and a drink (9)
16 Red bits sprinkled with love, quiet in there (10)
17 Hollow in tree (6)
18 With difficulty, getting clue for sultan (5)
20 Gatekeeper in street, desperately keeping quiet (2,5)
22 Provided expression to relegate first to last (6)
23 'Altogether' translated to Latin (2,5)
26 Louis in Rome as soldier followed him in Paris (5)
27 State has no right to this region (6)
28 See 33 Down
31 Not at all pleasant for a family to lose leaders (5)
33 Spot the instrument (not half) (3)
34 Premeditated murder of the oaf or thug (12)
37 Helping, in a modest way, to convert lisp to normal (5,7)
38 Divine naked woman's top half … (3)
39 … is seen at junction (approximately) (5)
41 Not so funny to lose therapeutic ends coming out of hospital (10)
43 Sucker for nothing (6)
45 Bleat about greeting in Brazil (5)
48 Fun with lute could be (7)
50 Got suited (6)
51 Legend rewritten about right monster (7)
53 A doctor's hesitant expression is a warning signal (5)
55 Heard where to find a filling that's not genetically modified (6)
56 Triumph of hope over experience to return stuff in anger, in anger (10)
61 Prince may be one that often makes the first move (5,4)
62 Firm stem (7)
63 Not one good open verse leads back to him (5)
64 See 20 Down
65 First atlas out to include 32 fjords won him an award (14)

Down

1 Beat the short, short gown (5)
2 Typical EU fiasco for Australian growers (9)
3 Whiskey's drunken bloomer (5)
4 Rogue state in charge of body (9)
5 Stops deer changing sides (5)
6 Ex-chancellor's make-up (4)
7 Exceeds the agreed limit of cricket terms (4,4)
9 Rude ripple in bum, shot in the arm (5,4)
10 Most dangerous start to driving here is across the channel (7)
11 African is up first to express disapproval (5)
12 See 1 Across
15 See 49
19 Stay calm as stand-up comedian larks about, not terribly amused (4,5)
20/40/36/64 14's farewell to Jesus from the multitude? (2,4,3,6,3,3,3,4)
21 Central to strict cycle, pick treatment for depression (9)
22 Same about wrong ratio of an imaginary line (10)
24 Arsonist gets cruel punishment at hearing (7)
25 Star, perhaps Paul McCartney, fluid in audition (10)
29 Kit quietly replaced by item of furniture — that's fair (9)
30 Try a wig: any style is a departure (5,4)
32 Homer may be one who says footnote is included (7)
33/35/28 President of the BBC and the French TV providing right coverage after strike carrier (6,10)
36 See 20
42 Spooner's baseball player at the restaurant (9)
44 Publication with earth a product here? (9)
46 Shrub that's hardy borders a garden that's informal (9)
47 Get stressed? Out of desserts? This will do it! (8)
49/15 Shoot missile round of 38's punishment (5,2,4)
52 Love prohibition in Scotland (4)
54 Judges reserve (5)
57 Keen on music, alternately out of tedium (5)
58 Fix hedge by trimming top (5)
59 Young partner for Max (5)
60 Where to have fun with blonde (4)

ACKNOWLEDGEMENTS

Emma, Nina, Sion, and Roger generously allowed me to describe their heart surgery experience. Tracey, Evgeny, surgeon H, and surgeon J generously provided eloquent and heartfelt narratives of difficult and vulnerable times. Peter Tallack of the Science Factory and Philip Gwyn Jones of Scribe recognised the potential of this book and supplied much encouragement. Ian Pindar did a superb job in editing the final manuscript and Molly Slight expertly steered it through to publication. To all of them, and to Fran for loving support and constructive critique throughout, thank you.